J.L. Cloudsley-Thompson · The Diversity of Amphibians and Reptiles

Springer

*Berlin*
*Heidelberg*
*New York*
*Barcelona*
*Hong Kong*
*London*
*Milan*
*Paris*
*Singapore*
*Tokyo*

John L. Cloudsley-Thompson

# The Diversity of Amphibians and Reptiles

## An Introduction

With 87 Figures

Springer

Prof. Dr. J.L. Cloudsley-Thompson
c/o Department of Biology
University College London
Gower Street
London, WC1E 6BT, UK

ISBN 3-540-65056-3 Springer-Verlag Berlin Heidelberg New York

Library of Congress Cataloging in-Publication Data
Cloudsley-Thompson, J.L. The diversity of amphibians and reptiles: an introduction / John L. Cloudsley-Thompson. p. cm. Includes bibliographical references (p. ) and index.
ISBN 3540650563 (alk. paper)
1. Reptiles. 2. Amphibians. I. Title.
QL641.C57 1999
597.9–ddc21

98-45640
CIP



Production: PRO EDIT GmbH, D-69126 Heidelberg
Cover design: design & production GmbH, D-69121 Heidelberg
Typesetting: Best-set Typesetter Ltd., Hong Kong

SPIN 10685072 31/3137 5 4 3 2 1 0 – Printed on acid free paper

## Acknowledgements

*I would like to extend my warmest thanks to my dearest wife Anne Cloudsley for her advice on a number of issues, to Mrs. Eileen Bergh for typing the manuscript, and to Dr. Dieter Czeschlik and Dr. Andrea Schlitzberger, Biology Editorial Springer-Verlag, for their friendly cooperation.*

# Contents


1 Introduction 1
2 Evolution 11
3 Convergence and Extinction 29
4 Locomotory Diversity on the Ground 41
5 Arboreal, Aerial and Aquatic Locomotion 61
6 Diversity of Anti-predator Devices 85
7 Nutritional Diversity 109
8 Reproductive Diversity and Life Histories 131
9 Daily and Seasonal Cycles, Hibernation, Aestivation and Migration 157
10 Thermal Diversity and Temperature Regulation 177
11 Water Balance and Excretion 197
12 Relationships with Mankind 213

Bibliography 227

Subject Index 243

# 1 Introduction

Despite current concern regarding the maintenance of biodiversity in the world, there is no universally accepted definition of the term 'diversity' and no single method of measuring it. In the present book, among other things, the variety of ways in which amphibians and reptiles have become adapted to live in different environments will be considered – that is, diversity of form and function – as well as the diversity of species that have evolved and co-exist within each particular environment or habitat. I have written primarily for students, amateur herpetologists, and all who are interested in natural history and wildlife generally. Let me, at the outset, therefore, reassure the reader both as to what will not be discussed and what will. To begin with the former, diversity is often regarded as a measure of the richness of species in a community or biome. This concept provides a more useful measurement of the characteristics of a community when it is combined with an assessment of the relative abundance of the various species present.

The index of diversity is a mathematical expression of the number of species, and the number of individuals of each species present. Various formulae may be used to quantify diversity. Of these, one of the most commonly employed is the Shannon-Wiener index. This is used when a community contains so many individuals that it can only be investigated by means of sampling. In this case, the index (D) is the ratio of the number of species to their importance as expressed, say, by their biomes or productivity within the community. $D = -\Sigma\, i^{s\ \pi} \log {}^{\pi}$; where $s$ = total number of species in the sample, $i$ = total number of individuals in one species, $\pi$ (a decimal fraction) is the number of one species in relation to the number of individuals in the population, and the log is to base $-2$ or base $-e$. The number of species in the environment is designated $\alpha$ diversity (intra-habitat), $\beta$ indicates the number of habitats or communities present (inter-habitat), and $\gamma$ diversity the rate of turnover.

Analyses such as this do not, however, take into account the variety of morphological forms present in a community or biome – terrestrial, cursorial, fossorial, aquatic, arboreal, aerial and so on. These are the topics with which this book is concerned. Moreover, related groups or taxa of smaller animals not only contain more species than do groups of larger ones living in the same habitat, but also contain more individuals of each species. This is simply because, on a small scale, there are many more micro-environments in the same region, e.g. areas of shrubs, trees, grasses, rocks, and so on, than there are in major habitats. The smaller the scale, the greater is the number of micro-habitats and, consequently, of species and individuals. Furthermore, it is a matter of opinion whether $\alpha$ diversity should be reckoned in terms of the number of species present or of the genetic diversity within each species (see May 1991).

More homogeneous environments must obviously contain fewer species than do less homogeneous environments, and the most numerous species occupy the largest habitats. For example, the marine copepod crustacean *Calanus finmarchicus* has possibly one of the greatest biomasses of any animal species because it is adapted to pelagic life throughout the oceans. Similarly, the yellow-bellied sea snake (*Pelamis platurus*) is probably the world's most numerous reptile species, partly on account of its extensive marine distribution. At the same time, because life began in the sea, there are many more marine than terrestrial phyla, as the major groupings of animals are called.

The concept of diversity is so greatly complicated by the numerous criteria involved in its assessment that I have not attempted to define the term, any more than I would try to define precisely what is understood by the word 'species'. The reader will have to be content with a general idea as to what, in this book, is meant by both terms, because it is not possible to be more precise. Genetic diversity has not been considered in these pages, nor have the numbers of species in various biomes. I have, however, tried to give some impression of the general diversity of amphibians and reptiles with reference to the ways in which they have become adapted to life on earth. This approach may not, by some, be considered sufficiently rigorous. Nevertheless, other readers may find it more interesting than to consider genetic diversity among individuals within a population – as is currently fashionable in biological circles.

To some people, specialization is particularly rewarding; for others there is more interest in taking a generalized approach to biology. In most research it may be convenient to specialize – to investigate a rather narrow range of topics – but important discoveries are often made when generalized knowledge is focused onto a narrow subject, or when different specialists combine to attack a large problem, such as the chemistry of snake venoms. Both amphibians and reptiles are interesting not only in their own right and in the study of evolution – especially perhaps the fossil Mesozoic dinosaurs, pterodactyls, ichthyosaurs and plesiosaurs – but also as material for physiological and embryological research. Until comparatively recently, every student of biology was at some time in his or her career asked to compare the early stages of development of the fish, frog and fowl.

My own liking for amphibians and reptiles seems almost to be innate. I can just remember, at the age of five, collecting some animals from a pond and taking them to my father in part of a broken glass milk bottle. An older boy had told me they were 'tadpoles', but my father soon disillusioned me. He explained that they were mosquito pupae, and took me to another pond to catch real tadpoles. During my early childhood, and later at boarding school, I delighted in pond life. In those far-off days, ponds and streams were scattered throughout the English countryside. On holiday picnics with the family, and on free afternoons whilst at boarding school, I was able to indulge my hobby of collecting aquatic plants and animals for my aquarium and the garden pond. Amphibia were easy to come by – crested, smooth and palmate newts, frogs and toads – but reptiles were naturally less common, although, according to my diary, we found four adders, two of them black, within 100 yards of each other at Effingham, Surrey, on 6 May 1935. Grass snakes were more common. (I also had a pet Greek tortoise and a couple of European pond tortoises.) I mention this to indicate the extent to which amphibians and reptiles have decreased in Britain during the last 60 years or so. At the same time, there has been an alarming decline in amphibians, especially frogs, worldwide. This will be discussed in Chapter 12. On the other hand, it is very much easier to travel these days, and to visit tropical and subtropical countries where amphibians and reptiles are easily seen. Moreover, whereas before World War II comparatively few people in Europe or America were interested in amphibians and reptiles, today there are many herpetologists in both continents and elsewhere who possess extensive vivaria and maintain their captives in conditions suitable for breeding. This requires both skill and knowledge. These herpetologists are also frequently involved in conservation which is most beneficial to the environment generally.

At this point it may be advisable to clarify a few matters of nomenclature. First, the family Salamandridae (order Urodela* or Caudata) contains a large number of species, some of which are usually referred to as newts, others as salamanders. There are 12 species of newts in Europe and at least 10 other species that are known as salamanders. European newts are members of the genus *Triturus*: they tend to be less terrestrial than salamanders, but there are many exceptions to this generalization (see discussion in Griffiths 1996). Most North American Salamandridae are known as salamanders, but members of the genera *Taricha* and *Notophthalmus* are also sometimes called newts. Clearly, too much importance should not be placed on these common names.

Other English names that require explanation are turtle, tortoise and terrapin. In the United States all Chelonia are known as turtles, but in Britain this term is restricted to marine turtles. Terrestrial Testudinidae are called tortoises while the Emydidae, a large and varied assemblage of freshwater and semi-terrestrial turtles, are usually referred to as terrapins – although *Emys orbicularis* is sometimes known as the European pond tortoise! In general, throughout this book I have tended to use the term that is applied in the country in which the animal in question resides or by which it is known in the publications cited.

Some people are intimidated by the use of Latin names. This is unfortunate and unnecessary. It does not bother them to consult a telephone directory which is full of names, some of which will be unfamiliar but may be necessary for identification of the number one wants to phone. Generic names, the first of a pair of names in italics, are the equivalent of a surname. They are accorded a capital for the first letter, e.g. *Alligator*. The specific name follows the generic name and is spelled with a lower case first letter, e.g. *mississippiensis*. When the name is being used frequently the generic name may be abbreviated to its initial, e.g. *A. mississippiensis*, but this is only possible where no confusion is incurred. Family names, e.g. Alligatoridae, and subfamily names, e.g. Alligatorinae, should not be placed in italics. It is quite surprising how seldom these simple conventions are understood by the popular press! In this book I have usually given the Latin name (as well as the common name, where there is one) of each species in the hope that repetition will enable the reader to become familiar with the concept as painlessly as possible.

Herpetology, like arachnology, is a subject in which professionals and amateurs can meet on equal terms, at least as far as most field work is

---

* Caudata has priority, but Urodela is frequently used especially in older literature.

concerned. With current changes in the funding of biological research, the role of the amateur is becomming increasingly important. Indeed, at the dawn of the twenty-first century, much of classical zoology is reverting to the position that it occupied during the nineteenth century. This may not be altogether a bad thing, however, because there is so much knowledge and enthusiasm amongst the so-called amateurs. Only in certain specialized lines of research are these at a disadvantage in comparison with 'professionals', due mainly to the fact that they may not have equal access to technological innovations and, perhaps, because their biological knowledge is more restricted. So, it is partly to try to help in this second deficiency that, as a professional zoologist and a naturalist at heart, I have written the present book. I hope that it will help to broaden the outlook of the dedicated amateur, whose specialized knowledge probably greatly exceeds my own, encourage biology students to study herpetology and, most important, convey interesting information to all its readers, both amateur and professional, by bridging the gap between advanced tests and more popular introductions to herpetology. In addition, I have devoted somewhat less space to topics such as locomotion and reproduction, thermoregulation and water balance, which tend to be well covered in standard herpetological texts, but have given more attention to others, such as convergence and rhythmic activity, that usually receive less coverage. This bias of mine is reflected in the choice of titles included in the Bibliography.

To highlight the relationship between amateurs and professionals, I might mention that Malcolm Smith, author of *The British Amphibians and Reptiles* (1951), qualified as a Doctor of Medicine at Charing Cross Hospital, London, at the end of the nineteenth century. Soon afterwards, he began working in the Far East where he spent 25 years (mainly in Bangkok) where he was attached first to the British Legation and, later, to the Court of Siam. He had a particular interest in herpetology, and took part in a number of zoological expeditions before he retired from medical practice and began to work as a taxonomist at the British Museum (Natural History), now the Natural History Museum. Thus he was a naturalist who later became a professional herpetologist.

Amphibians and reptiles have sometimes played a role in research not directly involved with herpetology. Wilfred Neill (1974) explains how frogs were used in some of the earliest studies on electricity. Some 200 years ago an experimenter noticed that the legs of newly dead frog, hanging from a copper hook, would kick when they touched the iron railing to which the hook was attached. The observer assumed that electricity was coming from the frog's muscles, but a colleague claimed that it was coming from the two metals when the legs made a connection between them. This difference of opinion led to further investigations which proved, among other things, that the action of nerves on muscle is mediated electrically. Frogs were among the first experimental animals used in the space programme. In order to test the effect of zero gravity upon the sense of balance, microelectrodes were implanted in nerves leading from the inner ears of bullfrogs. When the frogs were sent into orbit, their nerves kept signalling at an abnormally high rate for 51 h, but then slowed down to the normal rate. They stayed at this level for the three remaining days of the experiment. This proved that the inner ear can adapt to weightlessness, a condition to which human astronauts likewise soon become adjusted.

Animal structures can often be explained best in terms of their function. For instance, the vertebral column of the frog has lost its primitive flexibility. There is one cervical or neck vertebra, seven trunk vertebrae and a sacral vertebra which supports the hind limbs. The large transverse processes articulate with the ilia of the pelvic girdle. Between them lies a long, central urostyle (Fig. 1). Shortening of the body, which can thus be more readily supported on the legs and is less prone to undulate in walking, is a characteristic feature of the change from aquatic to terrestrial life. The peculiar adaptations of frogs referred to above are, however, assocaited especially with the action of jumping. The frog is like an animated safety-pin; when the point is released from its housing, the two arms of the pin spring apart. This could not occur if the upper arm were replaced by a flexible chain, for then there would be nothing to provide the resistance against which the spring could react. Similarly, the frog would not be able to jump as it does, unless the posterior portion of its vertebral column were rigid.

The functions of the neural spines or spinous processes of the vertebrae can likewise be understood if the backbone, supported by two limbs at each end, is compared with a suspension bridge. As the spinal column of tetrapod or four-footed vertebrates supports the weight of the body, there is a tendency for it to sag in the middle. This is prevented by tendons and the muscles above and below the spine: these pull together the ends of the backbone which is therefore bowed upwards. They compress the spine and are themselves in a state of tension. Those attached to the neural spines also act like the cables of a suspension bridge. Throughout this book I have assumed a strict correspondence between structure and function. Although this may not necessarily be invariably true, it probably applies in all the examples discussed.

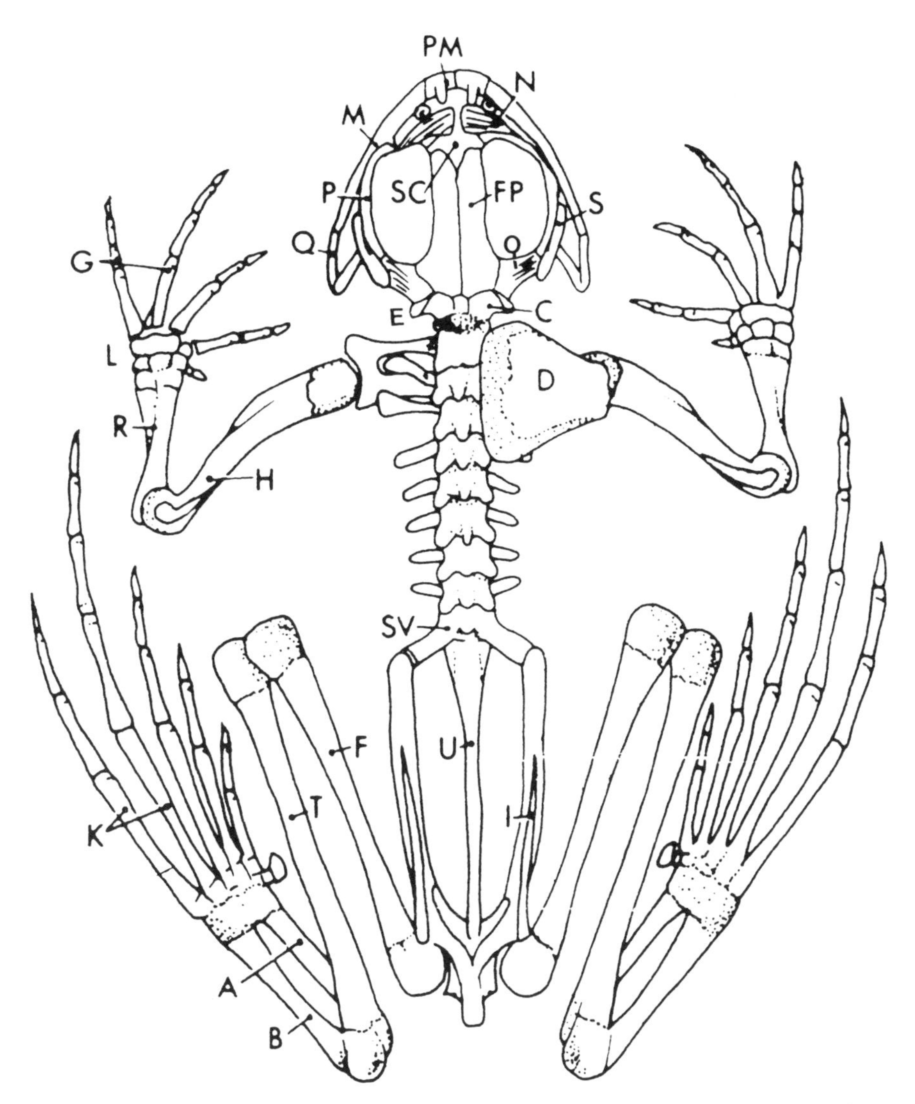


**Fig. 1.** Skeleton of the frog, seen from the dorsal aspect; the left suprascapula and scapula have been removed. *A* astragalus; *B* calcaneum; *C* columella auris; *D* suprascapula; *E* exoccipital; *F* femur; *FP* fronto-parietal; *G* metacarpals; *H* humerus; *I* ilium; *K* metatarsals; *L* carpals; *M* maxilla; *N* nasal; *O* pro-otic; *P* pterygoid; *PM* premaxilla; *Q* quadrato-jugal; *R* radio-ulna; *S* squamosal; *SC* sphenethmoid; *SV* sacral vertebra; *T* tibio-fibula; *U* urostyle. (Cloudsley-Thompson 1978a)

Most movements are achieved by the action of paired antagonistic muscles. A familiar example is afforded by the human biceps and triceps muscles (Fig. 2). When the biceps contracts the triceps relaxes and vice versa. Muscles cannot exert a thrust: after contracting, they can be restored to their original length only by the pull of another muscle attached to some skeletal structure. The two muscles of an antagonistic pair are unable to contract at the same time because the nervous excitation of one inhibits contraction of the other. Antagonistic muscles thus operate against one another indirectly through the skeleton of an animal.

Analogies may be both interesting and informative but should be interpreted with care. As Agnes Arber, in *The Mind and the Eye. A Study of the Biologist's Standpoint* (1954, Cambridge University Press, Combridge), emphasized, a frequent danger lies in the tendency for comparisons to become mechanical clichés. 'The warning note struck in Shakespeare's sonnet, "My mistress' eyes are nothing like the sun", is still as necessary as it was when it was uttered, for conventional language continues to belie everything which it touches "with false compare".' Analogies must avoid facile analogues, for instance, that individual senescence could be comparable with 'racial senescence'. This has been invoked as a possible explanation of the extinction of the dinosaurs. On the other hand, when we study parallel evolution it tells us

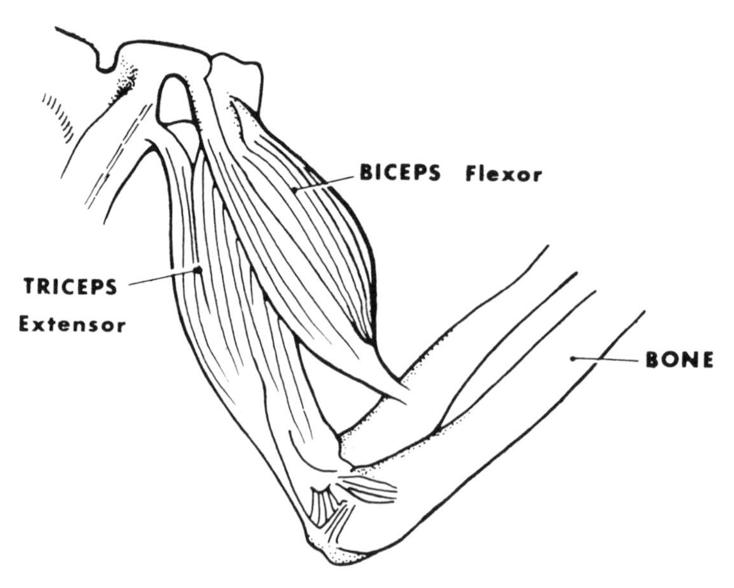


**Fig. 2.** Antagonistic muscles that move the human fore-arm. (Cloudsley-Thompson 1978a)

a great deal about the ways in which a particular type of environment or biome has engendered natural selection of the diverse groups of taxa of animals inhabiting it.

Diversity is not restricted to morphological characters alone. The adaptations of animals to different environments involve not only their structure but also their physiology and behaviour. It would scarcely benefit a crocodile, lurking in wait for its prey, if it were to leap about like a demented frog. Nor would there be an advantage to a scarlet arrow-poison frog to hide beneath the frond of a fern. Warning coloration has to be seen to be effective. Considerable diversity can be found in the physiology, nutrition, reproduction, thermal and water relations, and so on of amphibians and reptiles, and this diversity too requires consideration.

This book relies upon certain basic assumptions, foremost amongst which is that diversity is both a consequence of natural selection or the survival of the fittest and also of the basic material upon which natural selection operates. As in the game of chess, although there are limited rules or premises, the number of variations available within these is almost infinite.

# 2 Evolution

Life, almost certainly, began in the sea, and the first terrestrial vertebrates were amphibians. During the Upper Devonian period, the descendants of certain crossopterygian fishes ventured out of the water and began crawling from one pool to another. The Late Devonian was a time of increasing aridity, so that a selective advantage would have been obtained from being able to remain for a while on dry land. These first amphibians, or Ichtyostegalia, evolved in directions quite different from those followed by their ancestors. Although they had many characters in common, important differences appeared between the lung fishes and their allies and these early amphibians. This resulted from the specialization of the former for aquatic and, of the latter class, for terrestrial life. The Ichthyostegalia were four-limbed tetrapods and, while some were scaly, others had dry, leathery skins. Moreover, the ventral sides of the body were occasionally protected by bony plates. The skull of *Ichthyostega* shows the characteristic features of an amphibian whilst simultaneously retaining traces of its fish ancestry both in its shape and by the presence of a preopercular bone. In addition to the Ichthyostegalia (fish vertebrae), there were two other orders of Palaeozoic amphibians (Labyrinthodontia), viz. the Temnospondyli (divided vertebrae) and Anthracosauria (coal lizards). The reptiles evolved from Anthracosauria.

Throughout this book I have referred to Reptilia, although these animals are not today always regarded as a valid monophyletic group. Monophyly is the condition in which a group of taxa shares a common ancestry, being ultimately derived from a single interbreeding population, whereas polyphyletic groups are derived from many such populations. Reptiles are, nevertheless, united in the possession of a particular kind of life strategy – although there are even some exceptions to this.

Once established on land, both amphibians and reptiles diversified and became adapted to a variety of habitats. Some became specialized for climbing rocks and trees, some for burrowing in the soil or for flight, while others returned to life in the sea, or in freshwater. This is one of the kinds of diversity to be discussed in the present book. When several unrelated groups of animals become adapted to the same habitat we speak of it as 'convergence'. Seen laterally, as it were, this is 'parallel evolution' among the diverse groups or taxa. When considered from the viewpoint of the earliest forms, however, 'radiation' is observed to have taken place. Indeed, convergence, parallel evolution and radiation are really aspects of the same phenomenon – adaptation – seen from different points of view.

Amphibians thus evolved from the Crossopterygii, a subclass of lobe-finned lung fishes. Even the most primitive crossopterygians possessed lungs; some also had internal nares (passages connecting the mouth to the exterior of the body); and their skeletons were well-ossified and bony, a prerequisite for life on land. The fossils of the earliest amphibians are always associated with freshwater deposits, and all the evidence suggests that they evolved from crossopterygian fishes of the order Rhipidistia which, likewise, inhabited shallow freshwaters during the Devonian and Carboniferous periods (Fig. 3). Both groups were carnivorous.

There is no general agreement about the classification and relationships of Late Palaeozoic amphibians. One series, the Lepospondyli, was characterized by the spool-shaped central portions of the vertebrae. It included the Aistopoda – limbless, elongated, snakelike forms. Although the earliest known in the fossil record, and almost certainly primitive to other Lepospondyli, the loss of legs is a specialized feature and their ancestors must have had two pairs of legs. A third group was the Nectridea. This followed two lines of evolutionary development, one characterized by an elongation of

**Fig. 3.** A tetrapod-like crossopterygian fish (*right*) and a primitive amphibian (*Ichthyostegia*) (*left*), both Devonian. (Based on Colbert 1965)

the body to an eel-like form (paralleling the Aistopoda), the other with bizarre horns formed by extension of the postero-lateral projections of the skull which caused it to be considerably wider than it was long (Fig. 4). All nectrideans had unusual, easily recognized lepospondylian-type vertebrae. The first of these two evolutionary trends is represented by the genus *Sauropleura*, which inhabited the swamps of the Carboniferous. *Diplocaulus*, from the Permian, probably represents the culmination of the second trend. Species of the genus were evidently water-living amphibians which probably spent most of their time on the bottom of rivers and ponds. The functions of the horns are unknown. The Microsauria was another group of variable, but basically salamander-like animals which were probably not even lepospondylans. Indeed, they are sometimes grouped with the Lysorophia – semi-aquatic forms with reduced limbs (Romer and Parsons 1986).

The second series of Palaeozoic amphibians comprised the Labyrinthodontia (folded teeth) (Fig. 5) which were derived from the Ichthyostegalia of the Upper Devonian, the first terrestrial vertebrates, the crocodile-like Temnospondyli and the Anthracosauria – the ancestors of the reptiles.

**Fig. 4.** Reconstruction of *Diplocaulus*, a small, highly specialized and secondarily aquatic lepospondyl amphibian from the Permian. The lateral, horn-like, extremities to the skull might possibly have provided lift when swimming and assisted in burrowing into mud (length 50 cm)

**Fig. 5.** Reconstruction of *Eryops*, a large, Early Permian labyrinthodont amphibian. An effective competitor of contemporary reptiles, it inhabited freshwater and preyed on fishes (length 2 m)

The two major groups of extinct amphibians, the Lepospondyli and the Labyrinthodontia, are fully distinct when they first appear in the Late Palaeozoic fossil record. They were somewhat less well adapted to aquatic conditions than were the rhipidistian ancestors from which they evolved, yet the evidence indicates that they too were still primarily aquatic. According to some authorities, there was then a vast terrestrial niche available for vertebrates, while competition in water had become so severe that it placed a premium on the ability to be at least partially terrestrial. Another school of thought suggests that terrestrial adaptations may have evolved because they increased the probability for survival of their possessors not as terrestrial animals but in an aquatic environment. The survival of aquatic animals would have been enhanced during periods of drought if they could travel overland to reach larger and less crowded bodies of water at times when those in which they were living became greatly overcrowded with predators and competitors, or dried up (Porter 1972).

There are such close resemblances between the skulls of the Devonian crossopterygian fishes and the earliest labyrinthodont amphibians that their relationship is not in doubt, although there is no detailed fossil evidence of the stages of transition between the two types. While modern amphibians are generally considered to be a single monophyletic group – that is, as already explained, they are derived from a single common ancestor – their relationship to the Palaeozoic Amphibia is uncertain. The Rhipidistia occurred in two forms during the Devonian period, and the suggestion has been made that there may have been two distinct origins of the Amphibia, one from each of these.

Existing amphibians are the remnants of a dominant and abundant class of animals. This, coupled with large gaps in the fossil record, makes their classification and the determination of their relationships an extremely difficult problem. The ancestry of the worm-like Apoda* is not disclosed in the fossil record. The pedigree of the salamanders can be traced back to the Lower Cretaceous, but the one form known of that antiquity is already modern in its morphology and shows no connection with any more primitive type. Typical frogs, essentially modern in detail, are present in the Jurassic strata, and a genus, known as *Protobatrachus*, has been found in the Triassic rocks of Madagascar. It has a frog-like skull, but, although the backbone is somewhat shortened, it is considerably longer than that of a true frog and there is no development of the structure of the hind legs for hopping. Presumably the ultimate tetrapod ancestors were labyrinthodonts, but there are no definite links between the oldest salamanders and frogs and any known labryintodont forms. The small lepospondyls, mentioned above, differ from the labyrinthodonts in various features and might possibly be related to modern amphibians. At the present time, however, they cannot be linked with either labyrinthodonts or salamanders and frogs (Romer 1968).

Although they already possessed lungs, the ancestors of the modern Amphibia – the early Lissamphibia – must also have evolved an accessory respiratory surface. Scales became reduced and the moist skin of the body was used for this purpose. This can be deduced from the fact that, like modern amphibians, the ancestral lissamphibians did not expand their lungs by means of ribs, but forced air down their throats by means of special muscles and the hyoid apparatus. (The hyoid bone or bones support the floor of the mouth.) When the floor of the mouth of a frog or salamander is lowered, air is sucked in. The nostrils are then closed and the hypoglossal muscles force this air down to the lungs. This happens two or three times in quick succession, which is enough to fill the lungs. The throat muscles are then relaxed, allowing the air to be released from the lungs. (If the nostrils are

* Gymnophiona is technically correct, but Apoda is simpler and frequently used.

kept closed at this time, the air can be forced into the vocal sacs so that croaking occurs.) It seems unlikely that this method of breathing would have been adequate for an active land animal unless it were supplemented by accessory cutaneous respiration, as is also seen in existing amphibians (Young 1981).

No fossils have yet been found of the amphibian group immediately ancestral to reptiles, nor of the primitive reptilian stock prior to its initial diversification. The oldest fossils known are of the first mammal-like reptiles and of Anapsida. The latter are commonly regarded as being the more primitive because of their small size and unspecialized skeletons. They gave rise to the diapsids which include the majority of reptilian taxa, both fossil and extant. The origin of reptiles has been discussed by Carroll (1969a,b, 1988) and many other writers.

As in the case of Amphibia, which have declined in importance, living Reptilia are far less diverse than were their ancestors and represent only 4 of the 17 reptilian orders recognized by Romer (1966). Although the fossil record of reptiles is more complete than that of the Amphibia, there are still many questions to be answered regarding the inter-relationships of the various orders. A fundamental difference between primitive reptiles and their amphibian ancestors lies in the fact that amphibians produce anamniotic eggs whereas the eggs of reptiles possess an amnion, that is, an inner cellular layer surrounding a fluid-filled sac containing the embryo. The amniotic fluid provides the liquid environment that is necessary for an animal to develop on land. Now, there is seldom any direct evidence regarding the type of eggs produced by extinct forms, so it is almost impossible in many cases to determine which of the intermediate types of fossils are definitely amphibian and which represent reptiles. The most primitive forms that are definitely reptilian appeared in the Early Pennsylvanian subperiod of the Carboniferous period, but already there were at least two or three different phylogenetic lines whose relationships are as yet far from clear. It is, of course, certain that reptiles did evolve from Amphibia, but the evolution of amniotic eggs may well have occurred *before* the Reptilia became fully terrestrial. The original reptiles, like modern turtles, could have been aquatic animals which deposited their eggs on land. (On the other hand, the adult *Seymouria* was a terrestrial amphibian with many reptilian characters, yet it had aquatic larvae.) As in the case of modern tropical frogs which oviposit on land in nests of foam, as we shall see, most of their predators were aquatic, so there would have been a selective advantage in depositing eggs on land where they were relatively safe (Porter 1972).

By the end of the Pennsylvanian, there were at least two major phylogenetic lines, the pelycosaurs (order Pelycosauria) (Fig. 6) which eventually gave rise to the therapsids and then the mammals, and the Cotylosauria (Fig. 7). The latter were primitive, unspecialized reptiles, probably ancestral to all other reptilian groups as well as to the birds. For this reason, they are known as 'stem reptiles'.

Although most herpetologists know that the dinosaurs were probably warm-blooded, they are not always aware that the mammal-like reptiles or Therapsida (Fig. 8) and their predecessors, from which the earliest mammals evolved, may well have been homeothermic also. The world of therapsid reptiles, from Middle Permian to Late Triassic times, about 240–190 my B.P. (million years before present), was one of profound changes in the earth's crust and of great diversification of terrestrial vertebrates. Although the land masses remained joined together to form a single continent Pangaea, many ranges of mountains appeared, and there was a general reduction in the extent of inland waters which, by the Late Triassic, had shrunk to roughly half their original size. Endothermy or the production of metabolic heat sufficient to maintain a constant body temperature was undoubtedly an important factor in therapsid evolution.

**Fig. 6.** Reconstruction of *Dimetrodon*, a Lower Permian pelycosaur. The 'sail' on its back was a thermoregulatory device (length 3 m)

**Fig. 7.** Reconstruction of *Diadectes*, a heavily built cotylosaur of the Permian. The dark spots are imaginery (length 1.5–2 m)

**Fig. 8.** Reconstruction of *Cynognathus*, an advanced theriodont therapsid of the Early Triassic. About the size of a large dog, it probably maintained a fairly constant body temperature and may well have had a coat of hair rather than scales

Therapsids themselves evolved from pelycosaurs during the Pennsylvanian subperiod, as we have seen, and enjoyed a dominant position among vertebrates until the Middle Permian. This was due to the fact that their jaw muscles were arranged in such a manner that struggling prey were not able to dislocate the lower jaw. The carnivorous pelycosaurs fed mainly on aquatic prey, whereas the therapsids, which supplanted them, had a cranial morphology associated with herbivory or predation on land (see discussion in Hotton et al. 1986).

The interrelationships of the various taxa discussed above can best be explained in the following table of classification (Table 1). This differs principally from that of Carroll (1988) in that the Chelonia are regarded as Anapsida and not as a separate subclass Testudinata.

**Table 1.** Classification of the Amphibia and Reptilia

| Classification | |
|---|---|
| Class Amphibia | |
| Subclass Labyrinthodontia (folded teeth) | |
| Order Ichthyostegalia (fish vertebrae) | |
| Order Temnospondyli (divided vertebrae) | |
| Order Acanthosauria (coal lizards) | |
| e.g. *Eryops* | (Fig. 5) |
| Subclass Lepospondyli (scale vertebrae) | |
| e.g. *Diplocaulus* | (Fig. 4) |
| Subclass Lissamphibia (smooth amphibians) | |
| Order Urodela or Caudata (with tails) | |
| Newts and salamanders | |
| Order Anura (without tails) | |
| Frogs and toads | |
| Order Gymnophiona (naked, serpent-like) or Apoda (without limbs) | |
| Caecilians | (Fig. 24) |
| Class Reptilia | |
| Subclass Anapsida (without arches) | |
| Order Cotylosauria (stem reptiles) | |
| e.g. *Diadectes* | (Fig. 7) |
| Order Chelonia (turtles) or Testudinata (tortoise-shelled) | |
| Turtles and tortoise | |
| Subclass Synaptosauria (connecting lizards) | |
| Plesiosaurs (ribbon lizards) (Fig. 36), Pliosaurs, Nothosaurs, etc. | |
| Subclass Ichthyopterygia (fish fins) | |
| Ichthyosaurs (fish lizards) | (Fig. 35) |
| Subclass Lepidosauria (scaly lizards) | |

**Table 1.** *Continued*

| | |
|---|---|
| Order Rhynchocephalia (beak-headed) | |
| e.g. *Sphenodon* | (Fig. 9) |
| Order Squamata (scaly ones) | |
| Lizards, snakes and amphisbaenians | |
| Subclass Archosauria (ruling lizards) | |
| Order Thecodontia (socketed teeth) | |
| Order Crocodylia (crocodiles) | |
| Crocodiles and alligators | |
| Order Saurischia (lizard pelvis) | |
| Theropod and sauropod dinosaurs | |
| Order Ornithischia (bird pelvis) | |
| Ornithopod dinosaurs, stegosaurs, ankylosaurs and ceratopsians | |
| Order Pterosauria (wing lizards) | |
| Pterodactyls, etc. | |
| Subclass Synapsida (jointed arches) | |
| Order Pelycosauria (Theromorpha, mammal-like) | |
| e.g. *Dimetrodon* | (Fig. 6) |
| Order Therapsida (mammal-arched) | |
| e.g. *Cynognathus* | (Fig. 8) |

(Some fossil orders, which have not been discussed much if at all in this book, have been omitted from Table 1.)

Knowledge of the interrelationships of reptiles depends mainly upon their fossil remains, of which skulls are the most important. In the subclass Anapsida, the dermal bones of the temporal region of the skull behind the eyes present an unbroken surface, and there are no apertures between them. This characteristic separates the cotylosaurs and chelonians from all other reptiles. It is a primitive condition resembling that found among the early amphibians. In more advanced reptiles there are holes or fossae surrounded by bony arches in the temple region. These allow the jaw muscles to bulge into them, thereby facilitating their action. In the Lepidosauria and Archosauria, probably the most successful subclasses of reptiles, there are two such fossae and the condition is termed 'diapsid'. The lower temporal arch is incomplete in the Squamata, while the upper arch, too, is lost in some lizards and in snakes (Fig. 10).

Only a single fossa and arch is present in the remaining subclasses. When these are situated high on the skull, the condition is known as 'parapsid'. Both the Ichthyopterygia (ichthyosaurs) and Synaptosauria (plesiosaurs, etc.) have parapsid skulls and, at one time, both were included in a single subclass

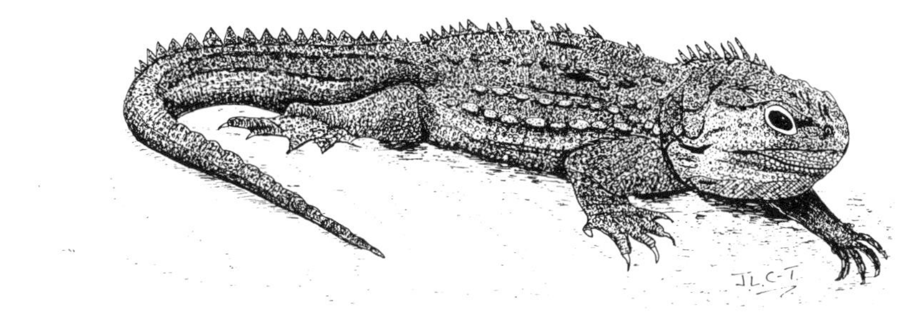

Fig. 9. *Sphenodon*, the New Zealand tuatara (length ca. 75 cm). (Cloudsley-Thompson 1994)

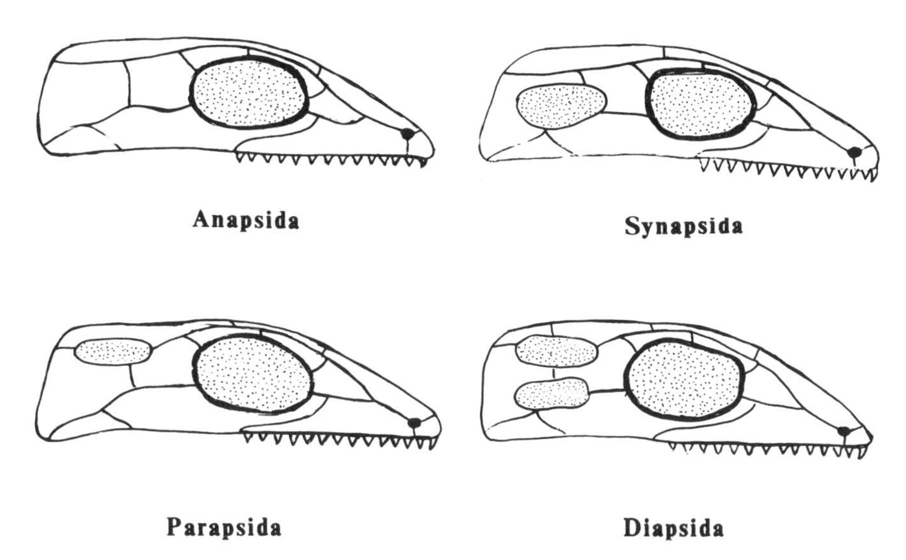


Fig. 10. Reptilian skulls, showing arrangement of the temporal openings. The various bones have not been named. (Based on Young 1981)

Parapsida. It is now realized, however, that they are not closely related. In the remaining subclass, Synapsida, there is also a single fossa but, in the earlier forms at least, this was located lower in the skull (Fig. 10). The term 'synapsid' means 'fused arch' and was originally given by early workers who believed, incorrectly, that the single arch had been derived from the fusion of the two arches found in diapsid reptiles (see discussion in Young 1981). From these remarks, it will be appreciated that Table 1 reflects basic differences between the various subclasses of reptiles. These began to appear during the Carboniferous and Permian periods and reached their zenith during the Mesozoic era (Triassic, Jurassic and Cretaceous periods). All existing Amphibia belong to a single subclass Lissamphibia, as mentioned above, but modern reptiles are allocated to three subclasses: Anapsida, Lepidosauria and Archosauria. The Lepidosauria contains two orders with living representatives. The Rhynchocephalia comprises the tuataras of New Zealand, which have survived almost unchanged since the Late Permian, whilst lizards, snakes and amphisbaenians are all included in the Squamata.

The Archosauria were the dominant land animals of the Late Mesozoic, and included the dinosaurs and the pterosaurs. The crocodilians are their only living descendants, although birds are also derived from archosaurian stock. The first of the archosaurs were the Thecodontia - small carnivorous reptiles of the Triassic period with sharp teeth embedded in sockets along the edges of the jaws. Some of them ran on their hind legs like certain modern lizards and one group, the phytosaurs, became amphibious, showing remarkable parallelism with the Crocodylia. Two orders of archosaurs are included among the dinosaurs (terrible lizards). These dominated the land throughout the Mesozoic era, just as the ichthyosaurs and plesiosaurs dominated the waters of the oceans. The Saurischia comprised two suborders, Theropoda and Sauropoda. The earlier theropods were small, bipedal carnivores, but, by the end of the Cretaceous period, they had produced the largest carnivore, *Tyrannosaurus rex*, ever to have appeared on earth. The Sauropoda, on the other hand, were quadrupedal vegetarians culminating in the immense Jurassic dinosaurs *Apatosaurus, Brontosaurus* and *Diplodocus.*

The second order of dinosaurs, the Ornithischia, appeared in the Jurassic and reached a peak in the Cretaceous, by which time the sauropods had already become less common. The first ornithischians were bipedal and included forms such as *Iguanodon*, but several lines reverted to a quadrupedal habit and some developed heavy armour and defensive spines (see Norman 1991; Cloudsley-Thompson 1994). The major difference between the two orders lies in the pelvis. In the Saurischia this is similar to that of other reptiles, including the thecodonts, while the pelvis of the Ornithischia is bird-like. In both groups, the hip socket is surrounded by three bones: the ilium above, the ischium below and behind, and the pubis below and in front. In the Saurichia the pubis projects forwards and downwards as a prong; in the Ornithischia it usually has two prongs, one projecting forwards and upwards, the other backwards so that it lies just below the ischium. In certain primitive ornithischians, the forward prong is absent (Fig. 11). Finally, the Triassic archosaurs gave rise to two independent taxa - the Pterosauria and the birds - both of which evolved wings and the ability to fly.

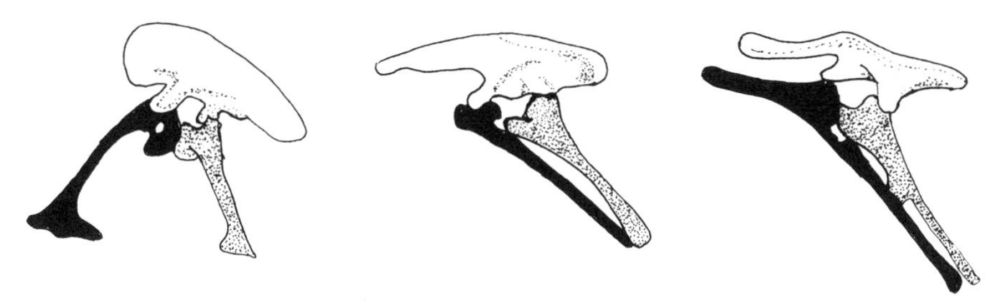

**Fig. 11.** Pelvic bones of dinosaurs. *Left* A saurischian; *centre* a primitive ornithischian; *right* a typical ornithischian. Ilium (*white*); ischium (*stippled*); pubis (*black*). The anterior is to the left in each instance

## 2.1 Size and Shape

One more factor of great evolutionary importance is the size of the animal in question. Differences in size make necessary changes in the structure and in the proportions of various parts of the body of an animal. If two organisms were of exactly the same shape so that one was a precise scale model of the other, they would be isometric in all respects. Every linear measurement of the larger would be $n$ times that of the smaller – the lengths and diameters of the body, limbs, internal organs and so on. However, the surface areas of all these parts, like that of the whole body of the larger organism, would be $n^2$ times that of the smaller, while its weight would be $n^3$. Conversely, the lengths and breadths of corresponding parts of isometric organisms would be proportional to (body weight)$^{0.33}$, areas to (body weight)$^{0.67}$, and volumes or weights equal to (body weight)$^{1.00}$. These simple relationships are responsible for the differences in shapes, structure and physiology between larger and smaller animals of the same groups or taxa. As we shall see in Chapters 10 and 11, size and size relations are extremely important in thermal and water relations.

The size of any animal is limited by its structural design which, in turn, is related to its mode of life. The largest of the sauropod dinosaurs probably weighed some 85 tonnes; the African elephant, the largest living land animal, weighs up to 10 tonnes, while the largest of all prehistoric land mammals, a long-necked hornless rhinoceros found in Asia between 25 and 35 my B.P., probably weighed about 15.5 tonnes. After a certain size has been reached, life on land becomes impossible for animals. This is because the strength of the legs and the muscles which support the body is principally a function of their cross sections. The cross-sectional areas of corresponding bones and muscles in isometric animals would be proportional to (body weight)$^{0.67}$, as we have seen, while the stresses due to body weight would be proportional to (body weight)$^{0.33}$. Large animals are made of the same materials, however, as their smaller relatives, and the stresses these materials can withstand are the same. A very large terrestrial animal must therefore have disproportionately thick limb bones and muscles, and stand and walk in such a way as to minimize stresses. This can be seen in the weight-bearing limbs of the dinosaurs, elephants, rhinoceroses and other large terrestrial vertebrates.

This problem does not arise in aquatic animals, the weight of which is supported by buoyancy, so that the larger whales can be considerably heavier than any land animals that have ever lived. There is a limit, however, even to the sizes of animals whose weight is supported by seawater. A point arises at which organisms can no longer deal with essential metabolic processes. They enter a vicious circle: larger bodies require bigger lungs, gills, intestines, kidneys and so on, all of which become relatively smaller with increases in linear dimensions, until a point is reached at which a large animal can no longer compete successfully with smaller ones of the same kind. It is highly improbable that any animal actually reaches the physical limitations inherent in its structural design, as it would lose its ability to compete successfully before then. Selection operates not on single parameters but on combinations of sometimes incompatible factors.

Natural selection can be defined as the survival of the most fit, with the inheritance of the factors in which the fitness lies. Successful individuals are those that survive long enough to be able to reproduce and bequeath genetic material to their offspring and subsequent generations. Even if an animal were to become so large and powerful as to be invulnerable to predatory enemies, it would not necessarily be successful. If success is to be evaluated in terms of passing on selfish genes to the next generation, it might be more effective to produce a larger number of young which reach maturity more rapidly and with less expenditure of metabolic energy in so doing. Natural selection must at times favour economy, even at the expense of survival and security.

The whale shark has the largest eggs of any living animal: they may be up to 28 cm long. Ostrich eggs measure 17 × 13.5 cm and weigh 1400 g, but the largest dinosaur eggs known are only 30.5 × 25.5 cm – smaller even than those of the extinct elephant bird of Madagascar, which measured 33 × 24 cm. The size of an egg is limited by the fact that, although the shell may be hardened to protect the embryo developing within, it has to be sufficiently porous to

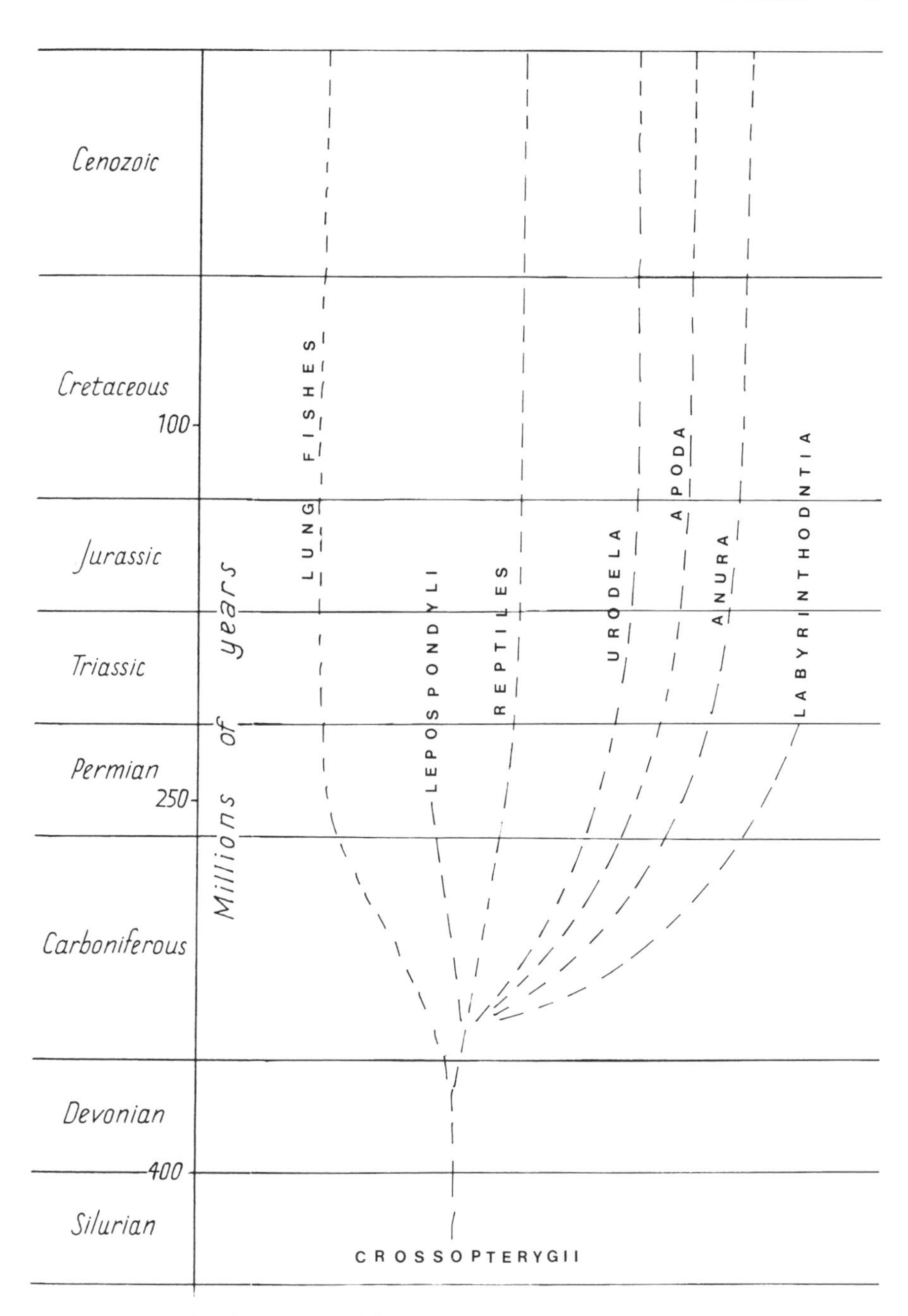


**Fig. 12.** Course of evolution in Amphibia. (Based on Young 1981)

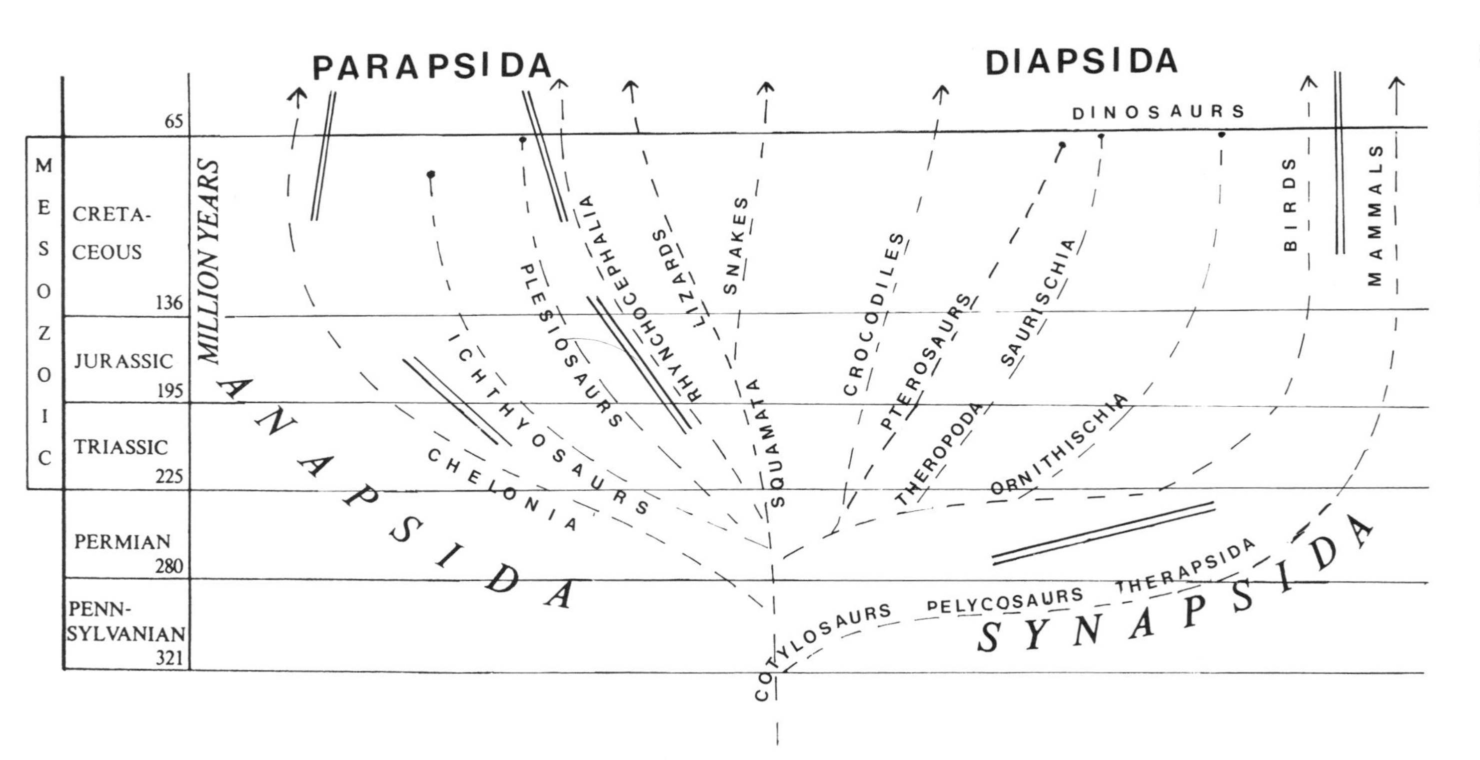


**Fig. 13.** Course of evolution in Reptilia. (Based on Young 1981)

permit the passage of air. It cannot, therefore, be inordinately thick or it will prevent respiration of the embryo from taking place. Moreover, it cannot be too hard or it will prevent the young animal from hatching. Since the thickness of eggshells is limited in this way, there is a geometric limit to the sizes of eggs. This limit is approached when eggs reach such a volume that the pressure of the internal fluid exceeds the strength of the shell that contains it. Consequently, the sizes of eggs are limited both by physical considerations and the fact that it may be more profitable to an animal to lay a larger number of smaller eggs even though the young hatch in a less advanced state of development (see Chapter 8).

Short-lived animals that produce numerous offspring are said to be *r*-strategists. Such ephemeral species contrast with K-strategists which invest heavily in a smaller number of young. These hatch, or are born, in a more advanced stage of development, but are cared for and defended by their parents. K-selected species tend to enjoy longer lives and attain greater size than do *r*-selected species. These are, however, the two extremes of a spectrum. Most amphibians are *r*-selected, with the exception of the Surinam toad (*Pipa pipa*), the South American marsupial frog (*Gastrotheca marsupiata*) and other species that show parental care, while crocodiles and pythons are K-selected. So, probably, were many of the dinosaurs. Chelonians are K-selected insofar as long life and relatively large size are concerned, but they are *r*-selected in that they lay many eggs and show no parental care of their offspring.

This brief account of the evolution of amphibians (Fig. 12) and reptiles (Fig. 13) has inevitably resulted in gross oversimplification. Hopefully, some of the defects on this score will be remedied in succeeding chapters of this book. For further information, however, the reader is referred to the works of Colbert (1962, 1965), Romer (1966, 1968), Schmalhausen (1968), Carroll (1969a,b) and Young (1981) (see also Olson 1976; Tweedie 1977; Norman 1985, 1991).

# 3 Convergence and Extinction

The more diverse a habitat, as we have seen, the larger the number of species that are likely to be found within it. Furthermore, different kinds of environment are inhabited by animals of different shapes and sizes. Subterranean burrowers, such as legless amphibians (Gymnophiona or Apoda, apart from one aquatic species), as well as amphisbaenians, slow-worms and some snakes, for instance, are quite dissimilar in appearance from aquatic forms, which may either have webbed digits, like frogs, turtles and plesiosaurs, or are finned and swim with their tails in the manner of fishes. Newts and salamanders, ichthyosaurs and sea snakes provide examples of the latter. Animals may also become adapted to the same environment in different ways. The diversity of amphibians and reptiles lies not so much in their diversity of form as in the number of times that a particular shape has evolved independently in different taxa. So far, I have mentioned only morphological adaptations, but animals are also adapted to their environments both physiologically and behaviourally. These topics will be discussed in later chapters.

## 3.1 Convergent Evolution

When unrelated species come to look alike as a result of parallel evolution in different geographical regions, they may be known as 'ecological equivalents' or, more precisely, as 'ecological analogues'. Some of the best known examples are to be found in the desert ecobiome. Here, the extreme physical and climatic conditions have engendered or enhanced a number of interrelated morphological, behavioural and physiological adaptations, many of which are paralleled in various quite unrelated taxa of animals and even of plants (Cloudsley-Thompson 1991, 1996).

Some of the best known mammalian examples are the fennec fox of the Sahara, which has a number of adaptive characters that parallel those of

**Fig. 14.** *Above Moloch horridus* (length ca. 10 cm); *below Phrynosoma platyrhinos* (length ca. 10 cm). (Cloudsley-Thompson 1991, 1994)

Nearctic kit foxes, and the kangaroo rats of the North American deserts, which closely resemble the jerboas of the Palaearctic realm. Reptilian examples include the Australian thorny devil *Moloch horridus* (Agamidae) and its approximate analogues *Phrynosoma* spp. (Iguanidae), the horned lizards of the North American deserts: all of these exploit a diet of ants (Fig. 14). No lizard of any other desert region of the world has adopted a similar lifestyle (Pianka 1986).

Another well-known reptilian example of ecological analogues is provided by the North American sidewinder (*Crotalus cerastes*) (Crotalinae) and its Palaearctic counterpart, the Saharan sand viper (*Cerastes cerastes*) (Viperinae) (Fig. 15). Both these snakes move by throwing lateral loops forward, and hide themselves by flattening their bodies and shovelling sand over their backs. They are so much alike in general appearance that, were it not for the rattle of *Crotalus cerastes* and the pits between its eyes and nostrils, the two might easily be confused (Cloudsley-Thompson 1991).

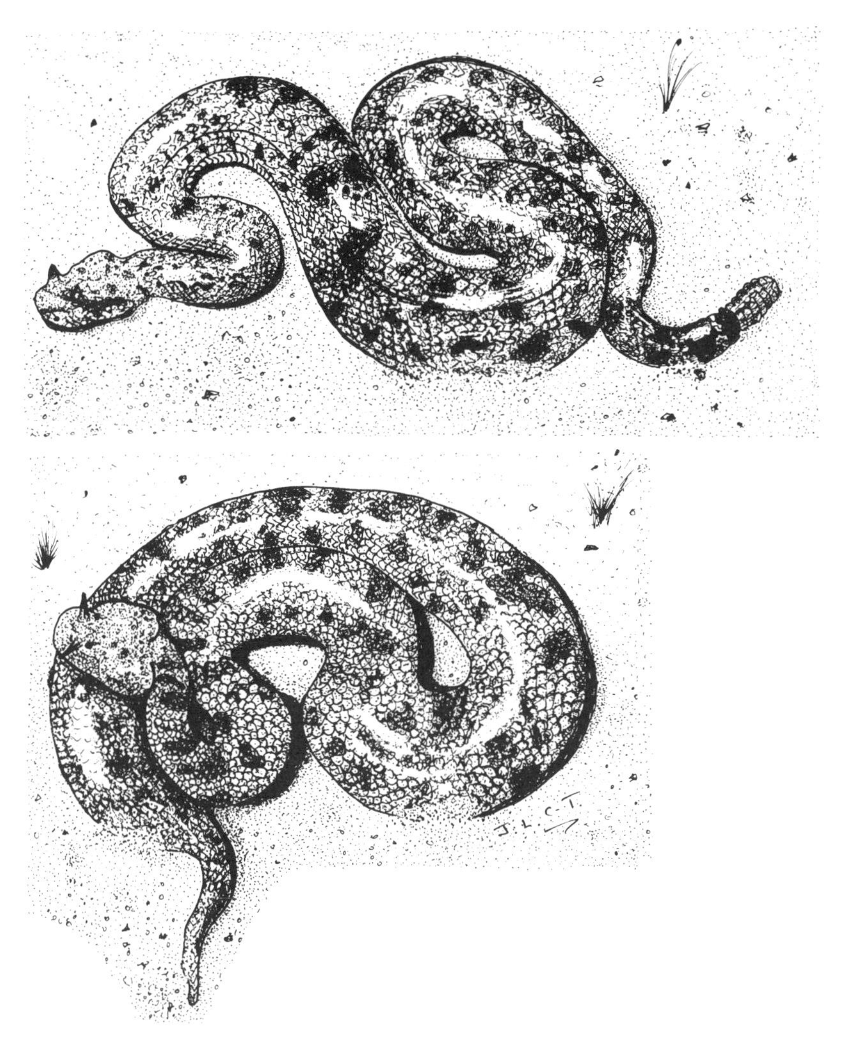

**Fig. 15.** *Above Crotalus cerastes* (length ca. 75 cm); *below Cerastes cerastes* (length ca. 75 cm). (Cloudsley-Thompson 1991, 1994)

In his analysis of the ecological niche and comunity structure of the lizard faunas of the North American, Kalahari and Australian deserts, Pianka (1986) pointed out that both North America and Australia have long-legged species that frequent the open spaces between plants – the iguanid *Calisaurus draconoides* and *Amphibolurus* (= *Ctenophorus*) spp. (Agamidae) respectively – while each region has a medium-sized lizard-eating species – *Crotaphytus wislizeni* (Iguanidae) in North America and *Varanus eremius* (Varanidae) in Australia. A few Kalahari–Australia species pairs are also roughly equivalent, e.g. the subterranean skinks *Typhlosaurus* and *Lerista* spp. (Scincidae) and the semi-arboreal *Agama hispida* and *Amphibolurus* (= *Pogona*) *minor* (Agamidae).

In addition to the examples already mentioned, other members of the Agamidae frequently occupy similar or analogous ecological niches to those of Iguanidae. Examples include *Uromastyx* spp. (Agamidae) of the Great Palaearctic desert and North American *Sauromalus* and *Cachryx* spp. (Iguanidae). The Southeast Asian agamids *Leiolepis* spp. resemble North American iguanids such as *Dipsosaurus dorsalis* and the Australian *Amphibolurus pictus* (Agamidae). The Iranian agamid *Phrynocephalus mystaceus* and the African gecko *Geckonia chazaliae* may also have many characters in common. Despite this, Pianka (1986) concluded that the differences between the ecologies of most lizard species in the three continental deserts that he studied are much more striking than are the similarities. 'It is easy to make too much out of convergence, and one must always be wary of imposing it upon the system under consideration.'

Nevertheless, recognition of convergence is an important factor in the understanding of natural selection. (Another example of convergence is provided by the adaptive coloration of desert animals, including reptiles, which nearly always match the sandy hues of their environment, or else are black.) It is almost a tautology to state that a particular biome, such as the desert, should engender comparable adaptations in its fauna in different zoogeographical realms of the world. Pianka (1986) provided an excellent example based on scorpion predation. While scorpions are solitary prey items, they are extremely large and nutritious, thereby presumably facilitating the evolution of dietary specialization. In the Kalahari they are preyed on by *Nucras tessellata* (Lacertidae) and in Australia by *Pygopus nigriceps* (Pygopodidae). The diurnal *N. tessellata* forages widely to capture these animals in their daytime retreats, whereas the noctural *P. nigriceps* sits and waits for scorpions moving at night, aboveground, during their normal period of activity. No North American desert lizard specializes on a diet of scorpions, but the small snake *Chionactis occipitalis* (Colubridae) appears to have usurped this ecological role.

Whereas some lizards have evolved as dietary specialists, rather more are generalists. *Moloch horridus* and *Phrynosoma* spp. mainly eat ants. The Kalahari lizards *Mesalina lugubris* (Lacertidae) and *Typhlosaurus* spp. (Scincidae) feed entirely on termites, as do the Australian nocturnal geckos *Diplodactylus conspicilatus* and *Rhynchoedura* spp. as well as some day-active *Ctenotus* spp. (Scincidae) (Pianka 1986). Dietary diversity occurs in many species of lizards which eat almost everything they can catch and overcome. Variations in diet also occur within the same species, both from time to time and from place to place, as opportunities present themselves and the abundance of particular prey species fluctuates.

The existence of ecological analogues and convergence among reptiles is more evident in a comparatively homogeneous biome, such as desert, than it is in more complex environments. Nevertheless, it often becomes apparent in specialized ecological niches everywhere.

As already indicated, evolution can be considered from different angles. From the viewpoint of the taxon, it is seen as radiation. Thus, from the basic terrestrial reptilian stock have radiated aquatic, burrowing and aerial forms. When a particular environment such as the sea is considered, however, we see that different groups, including marine turtles, ichthyosaurs, plesiosaurs and sea snakes, have converged upon it, whilst viewed laterally, parallel evolution appears to have taken place. The terms covergence and parallel evolution are not restricted to a particular taxon, however. Although wings evolved only in insects, pterosaurs, birds and bats, animals that glide include flying fishes of various types, flying frogs, flying lizards, flying snakes, flying squirrels and flying phalangers (Fig. 16). Animals belonging to different classes, such as birds and bats, are not usually considered to be sufficiently alike to be regarded as ecological analogues; but the terminology is very imprecise and it is really only the concept that matters. A glance at the reconstructions of primitive amphibians and reptiles illustrated in the previous chapter shows their remarkable similarity in form. Indeed, some of the reptiles may well have usurped the ecological niches previously occupied by amphibians. The differences that are of evolutionary significance are apparent only when the bones of the skull are studied.

## 3.2 Extinction

Throughout the course of evolution, various groups of plants and animals have become extinct while others have persisted, thereby contributing to the diversity of forms alive at any one time. Furthermore, the members of taxa that do survive themselves evolve and change, giving rise to new taxa. There has long been a tendency to attribute the extinction of larger taxa, such as the dinosaurs, to specific factors. As a result of this, incorrect or misleading conclusions have not infrequently been reached. Many articles, books, radio and television programmes on the subject of extinction have been written by

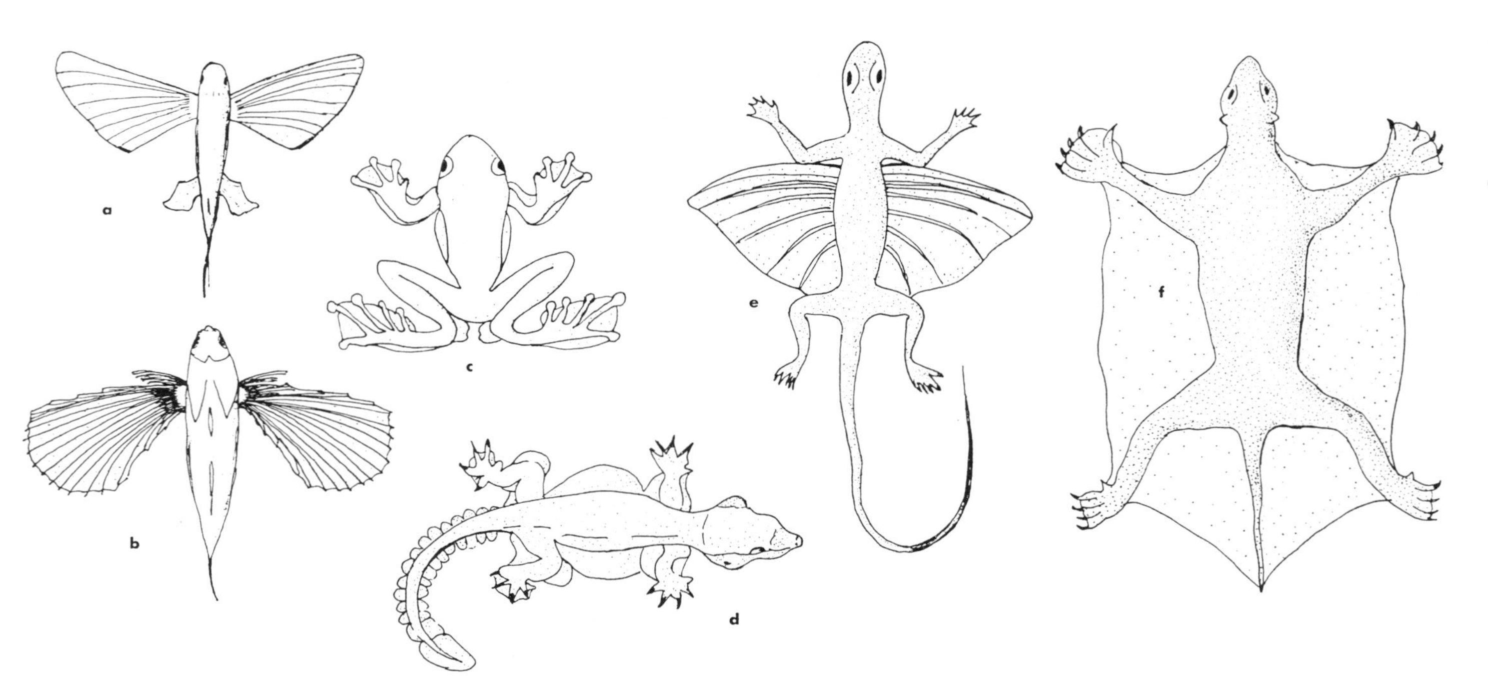


**Fig. 16.** Gliding animals. **a** Flying fish (*Exocoetus* sp.); **b** flying gurnard (*Dactylopterus volitans*); **c** flying frog (*Rhacophorus reinhardtii*); **d** flying gecko (*Ptychozoön* sp.); **e** flying dragon (*Draco volans*); **f** flying phalanger (*Galopithecus volans*). (Not to scale). (After various authors)

people who are not themselves palaeontologists, or even biologists. Secondly, ideas are often copied from sources that are outdated or incorrect. In addition, palaeontologists are often unsure of the events that led to the deaths of the animals that became fossilized (Charig 1979). Scientists, like everyone else, tend to believe their own hypotheses and to disregard those of other people with whom they disagree, although, in fact, both may be partially correct.

There have been five mass extinctions in the fossil record, the largest of these at the end of the Permian, the smallest at the close of the Mesozoic era – the Cretaceous–Tertiary (K–T) boundary, as it is called. (Geologists identify the Cretaceous period with a K rather than a C to distinguish it from the Cambrian and Carboniferous.) A mass extinction of amphibians had previously occurred, much earlier, at the end of the Permian period; this was the Appalachian revolution which saw the end of the Palaeozoic (ancient life) era (Fig. 12). Many other forms of life, especially among plants and invertebrates, also became extinct at this time. The invertebrates that survived have altered little until today, but the vertebrates, especially the reptiles, birds and mammals, have undergone practically their entire evolution since then. However, a large proportion of the dominant reptiles, both on land and in the sea, became extinct at the K–T boundary.

Of all mass extinctions, that of the dinosaurs and other ruling reptiles during the Laramide revolution at the end of the Cretaceous period has attracted the greatest popular interest (Wilford 1985). The groups that died out included the plesiosaurs, the theropod, sauropod and ornithopod dinosaurs, as well as the ankylosaurs; but the ichthyosaurs disappeared before the K–T boundary, as did the pterosaurs (Colbert 1962, 1965, 1980; Kurtén 1968; Swinton 1970; Young 1981; Behrensmeyer et al. 1992; Lambert 1992). This massive extinction is one of the greatest mysteries in the history of the world. Reptiles that had dominated the world for over 150 my vanished within a comparatively short period of time, as did marine invertebrate taxa such as the ammonites, various planktonic Foraminifera and rudist, reef-building bivalve mollusces.

In 1964, Glenn Jepson of Princeton University listed nearly 50 different hypotheses that had been proposed to explain the extinction of the dinosaurs: there have been many more since then (Croft 1982). These hypotheses fall into two broad categories: those that attribute extinction to a sudden catastrophic event, and those that propose that events over a longer period of time are responsible. A recent example of the former is the much publicized suggestion of L. W. Alvarez et al. (1980) and W. Alvarez et al. (1984) that an extraterrestrial bolide or asteroid struck the earth, plunging day into night, halting photosynthesis, and causing plants to become dormant or die. This hypothesis was based on the discovery of a layer rich in iridium, a rare element associated with extra-terrestrial bodies, in samples of clay from the K–T boundary in Italy. Iridium has subsequently been discovered in the K–T boundary in other parts of the world (Alvarez W. 1997). However, it still does not prove that this caused the extinction of the dinosaurs (Hallam and Wignall 1997).

A major problem that needs to be explained is how impact with a bolide even if, as it appears this did occur, could have selectively destroyed all the dinosaurs and plesiosaurs but spared other animals, such as the chelonians and crocodilians, some of which were extremely large. Furthermore, there are no iridium layers in the strata marking other major extinctions. Indeed, it has been suggested that the iridium anomaly may have been the consequence of volcanic activity and have nothing to do with an asteroid (McGowan 1991).

The alternative category of hypotheses attributes mass extinction to a combination of factors, of which climatic change is usually assumed to be the most significant (Halstead and Halstead 1981). These more synthetic ideas are able to encompass evidence from both palaentology and biology. They take into account the fact that the fall of the dinosaurs was not instantaneous; in North America the process took place over about 12 my.

Seasonal climates began with the break up of the original continent Pangaea and the beginning of continental drift. The summers became progressively hotter than the previous annual mean temperature, the winters cooler. In land tortoises, the heat produced by metabolic processes is matched by evaporative heat loss up to about 24 °C, but above this, additional cooling mechanisms are evoked to prevent overheating. I have argued that, for large reptiles, with correspondingly small surface to volume ratios, the hot summers were perhaps the one environmental change with which they were least able to cope (Cloudsley-Thompson 1978b). Semi-aquatic crocodilians would not have suffered to the same extent. On the other hand, if the end of the Mesozoic era was marked by desiccation so extensive as to have prevented the dinosaurs from doing the same, how did the crocodilians manage to survive? The Laramide revolution, with its uprising of land, nevertheless resulted in the draining of swamps, the acceleration of rivers and the gradual disappearance of everglades, which must have reduced crocodilian habitats. More recently, Milne (1991) argued that the world climate became cooler rather than warmer at the end of the Cretaceous, and that this was primarily responsible for the demise of the dinosaurs. I do not agree with this suggestion (see Chap. 10).

In 1972, Bakker proposed that the dinosaurs were endothermic homeotherms, more akin to birds and mammals than to most existing reptiles – a point taken up with enthusiasm by Desmond (1975). Desmond believed that the thinning of dinosaur eggshells and consequent sterility in the upper layers of the Upper Maestrichtian was the result of stress from cold. Croft (1982) postulated that the formation of cataracts in the eyes of elderly dinosaurs, due to excessive radiation, could well have been an important contributory factor to dinosaur extinction; while Bakker (1987) suggested faunal mixing might have resulted in the introduction of predators and diseases into populations which they had not previously encountered, thus resulting in the type of imbalance so often created by human agency. (For a discussion of the possible causes of the extinction of herbivorous dinosaurs see Chap. 10.)

At the end of the Mesozoic era, there was a great reduction in the abundance of marine plankton, and any proposed explanation of the mass extinctions that occurred then must take this into account. Phylotectonic maxima in the geological record coincided with, and probably caused, extensive biogenic deposition of calcium and evolutionary diversification of the contemporaneous fauna and flora. In contrast, decreased production triggered a general biotic turnover on both land and sea. Times of planktonic extinctions coincided with similar periods affecting animal taxa rather than terrestrial plants. These periods of severely reduced microfloras and low productivity resulted in depletion of atmospheric oxygen and an increase in the carbon dioxide content of the air, causing a 'greenhouse' effect and a consequent increase in global temperature. Thus, conditions during the Jurassic period apparently favoured the evolution of calcareous algae. The resulting algal bloom then depleted the atmosphere of carbon dioxide, and the sea of calcium and bicarbonate ions. In the deeper waters, a zone developed that was saturated with carbon dioxide. This moved to the surface during the K–T revolution, destroyed the phytoplankton and initiated an ecological collapse recorded by a gap of 1 my in the record of calcareous sedimentation (see Cloudsley-Thompson 1978b).

The plesiosaurs and mosasaurs died out during this period, but not marine turtles and crocodiles, the pterosaurs but not the archetypal birds. No doubt forms that disappeared were those less able to compete in a changing environment. I do not believe that mass extinction should be attributed to any single factor, although the hot summers of the seasonal climate that developed at the close of the Mesozoic era are probably the most important factor. Extinction would have been the outcome of a combination of unfavourable factors, including interspecific competition, always assuming that the plesiosaurs and dinosaurs did indeed die out. According to Ostrom (1976), however, the more lightly built theropod saurischians were the immediate ancestors of birds. Ostrom pointed out that, with few exceptions, virtually every skeletal feature of *Archaeopteryx lithographica* can be found in contemporaneous and near-contemporary coelurosaurian dinosaurs, and that many of these conditions are unrelated, specialized characters. This, he claimed, established that the closest ancestral affinities of *Archaeopteryx* are with coelurosaurian theropods. There is no contrary evidence, and 'any other explanation is illogical'. Avian phylogeny can therefore be represented as follows: Pseudosuchia – Coelurosauria – *Archaeopteryx* – higher birds (Ostrom 1976). If this view is accepted, then, of course, the dinosaurs are by no means extinct. The matter is, however, still the subject of dispute (Charig 1979). For instance Feduccia (1980, 1996) argues that birds are not dinosaurs. At least they cannot be placed among the late theropods such as the dromaeosaurs. The most bird-like dinosaurs are Cretaceous, whereas the birds were well established by the Jurassic period. Moreover, probably many of the theropods evolved feathers which, therefore, are not unique to birds.

Birds flourished during the Mesozoic era and colonized numerous ecological niches. The number of recognized fossil bird taxa has more than doubled during the last 5 or 6 years, shedding light on the large temporal and evolutionary gap between Late Jurassic *Archaeopteryx* on the one hand and superficially diver-like hesperornithiform and pelican-like ichthyornithiform birds on the other. The new findings provide additional evidence for the hypothesis that birds evolved from cursorial, bipedal carnivorous dinosaurs. The knee flexion and tibio-tarsal displacement of modern birds were acquired only late in avian history and differ from the ancestral theropod pattern of hip extension, which involved extensive femoral retraction during each stride, as is found in crocodiles and lizards today. Reduction of the tail and caudal musculative throughout theropod evolution is probably the principal agent involved in this mechanism, correlated with the functional decoupling of the tail from its primitive state, and its junction instead with the flight apparatus, plus the forward migration of the centre of gravity and the modern avian stance (Chiappe 1995).

The palaeontological evidence that different groups of dinosaurs died out at different times, sometimes surviving whatever it was that produced the iridium anomaly of L. W. Alvarez et al. (1980) and W. Alvarez et al. (1984), is indisputable. There was a great decline in the diversity of dinosaurs at the K–T boundary extending over the last 5 my of the Cretaceous period. The final 300 000 years witnessed an acceleration of this decline so that, in the end, there were only about 12 species in 8 genera left to become extinct (Charig 1989). Clearly, conditions were becoming unfavourable for the large terrestrial reptiles.

It is usually difficult and often almost impossible to prove that something does not exist. Most zoologists are properly sceptical about the possibilities of discovering any more new species of large animals. Nevertheless, there have been numerous, usually unsubstantiated, accounts in the literature of cryptozoology of unknown animals that might possibly be dinosaurs, plesiosaurs, mosasaurs or even pterodactyls (e.g. Heuvelmans 1959, 1968; Mackal 1983; Shuker 1995). Many of the reported sightings of lake monsters could be attributable to atmospheric distortion of the image so that familiar objects, such as sticks and branches, can easily take on unrecognizable forms. The majority of sightings are from elevations close to the level of the water surface with the observer near to the shore or in a boat – and horizontal light rays are easily deflected by refractile anomalies in the air! The type of motion described in many reports is also consistent with the observation of refractive events. Within a stationary inversion layer, the nature of a transmitted image is sensitive to variations in the elevation of an observer: it can readily undergo a large vertical shift in response to a small vertical movement of the observer. Moreover, under centain conditions, inanimate objects may appear to be moving even when the observer is stationary. If the inversion layer is in slow motion, the image may grow, shrink, move about sinuously, disappear and then reappear without a sound or ripple (Lehn 1979).

No doubt most reported sightings of dinosaurs and other monsters are likewise cases of mistaken identity. During the Middle Ages people used to see witches riding on broomsticks. In the twentieth century these have been replaced by flying saucers! However, can mistaken identity account for everything? Shuker (1995) and others argue very persuasively that it cannot. The thought that examples of giant Mesozoic reptiles might still exist, almost undetected, in the more inaccessible regions of the world cannot fail to excite. Yet it is somewhat surprising, if any of these do really occur, that there should be no uncontroversial Cenozoic fossils of the taxa concerned. Nevertheless, this point can readily be explained by some cryptozoologists. Until tangible evidence has been produced, it will probably be advisable to keep a fairly open mind on the subject and not to reach definite conclusions either one way or the other. As I suggested earlier, extinction is probably the result of a combination of multiple causes; and taxa die out when they can no longer compete effectively with others in a changing environment (see Cloudsley-Thompson 1978b).

# 4 Locomotory Diversity on the Ground

The earliest terrestrial vertebrates had short, stumpy legs and walked on the surface of the ground: burrowing, arboreal and aerial forms evolved from them, and will be discussed later. The first of the amphibians included a variety of types of which the order Ichthyostegalia of the Upper Devonian is best represented in the fossil record. These labyrinthodonts probably evolved from the Rhipidistia, as we saw in Chapter 2, although gaps in the record make it impossible to determine the exact line of descent. Even today, most amphibians and reptiles walk upon four legs as did their primitive ancestors, but modern forms usually move more quickly in proportion to their sizes. (For details of the locomotory processes of vertebrates in general, and of amphibians and reptiles in particular, see Gray 1968; Bellairs 1969; Young 1981; Little 1983; Duellman and Trueb 1986; McGowan 1991.)

## 4.1 Quadrupedal Locomotion

Of today's tetrapods, the Urodela or Caudata (newts and salamanders) show the most primitive locomotory traits. Their bodies are essentially fish-like in shape, and three different methods of locomotion can be distinguished on land. When they are frightened and need to move fast, they wriggle along with the belly resting on the ground, as though they were swimming on land, with their legs scarcely touching the soil. On the other hand, when moving deliberately, they carry their weight clear of the ground on their four legs. This places a completely different set of stresses on the vertical column which has then to act as a girder carrying the weight of the body and transmitting it to the legs. (In fishes, the spinal column merely bends from side to side in swimming.) The movement of one fore limb of a tetrapod vertebrate is followed by that of the hind limb on the opposite side. The fore leg on that side is then moved, followed by the rear leg on the opposite side. (It is exactly the same when a human being crawls on hands and knees.) In this way, the body is always supported by three of the legs which form a stable triangle. A similar basic pattern of movement can be seen in all four-legged animals. As salamanders speed up, they begin to trot: then the two opposite limbs are moved together and, during most of each stride, there are only two feet on the ground at the same time. (The morphology and locomotion of amphibians is described in detail by Gray 1968; Duellman and Trueb 1986; Stebbins and Cohen 1995.)

In modern reptiles, such as lizards, which have elongated bodies and small, laterally projecting legs the method of locomotion is similar to that of urodeles. Correlated with similarity of movement is a general similarity in the plan of the skeleton. There are, however, certain advanced features which are characteristic of all reptiles. The head is usually carried above the level of the ground, and the neck is well developed. The first two cervical vertebrae are modified to form the atlas and axis bones, and all the vertebrae articulate with one another by a system of interlocking processes much more elaborate than that found in amphibians.

Reptile locomotory patterns on land may be affected by a number of factors, of which body weight is probably the most important. For instance, when out of water, the heavy green turtle (*Chelonia mydas*), like other large species of turtles, has to move in an unusual way, shifting its body forwards in a series of lurches by synchronous movement of all four limbs. The lighter young, however, progress by alternate movements of their paddles, as do smaller turtles and tortoises (Bellairs 1969).

Most living quadrupedal reptiles hold their legs out almost horizontally when they move, as did the majority of extinct forms. They walk with sinuous movements of the body, as do newts and salamanders (Fig. 17). The limbs describe wide, sweeping arcs in both effective and recovery movements, the humerus and femur moving in an approximately horizontal plane. When some lizards are running fast, the phase of the fore limb is ahead of that reached by the hind limb on the same side of the body, instead of after it, as in the normal tetrapod sequence. This also occurs in crocodilians: it may be due to the fact that the hind limbs are longer than the fore limbs. Consequently, one hind foot is placed on the substrate well in front of the point of contact of the fore foot on the same side of the body, and before the opposite fore foot has reached the ground. By the time it has done so, both limbs of the other diagonal have left the ground, and the lizard is momentarily bipedal with only one rear leg touching the substrate. (True bipedality in reptiles is discussed below.)

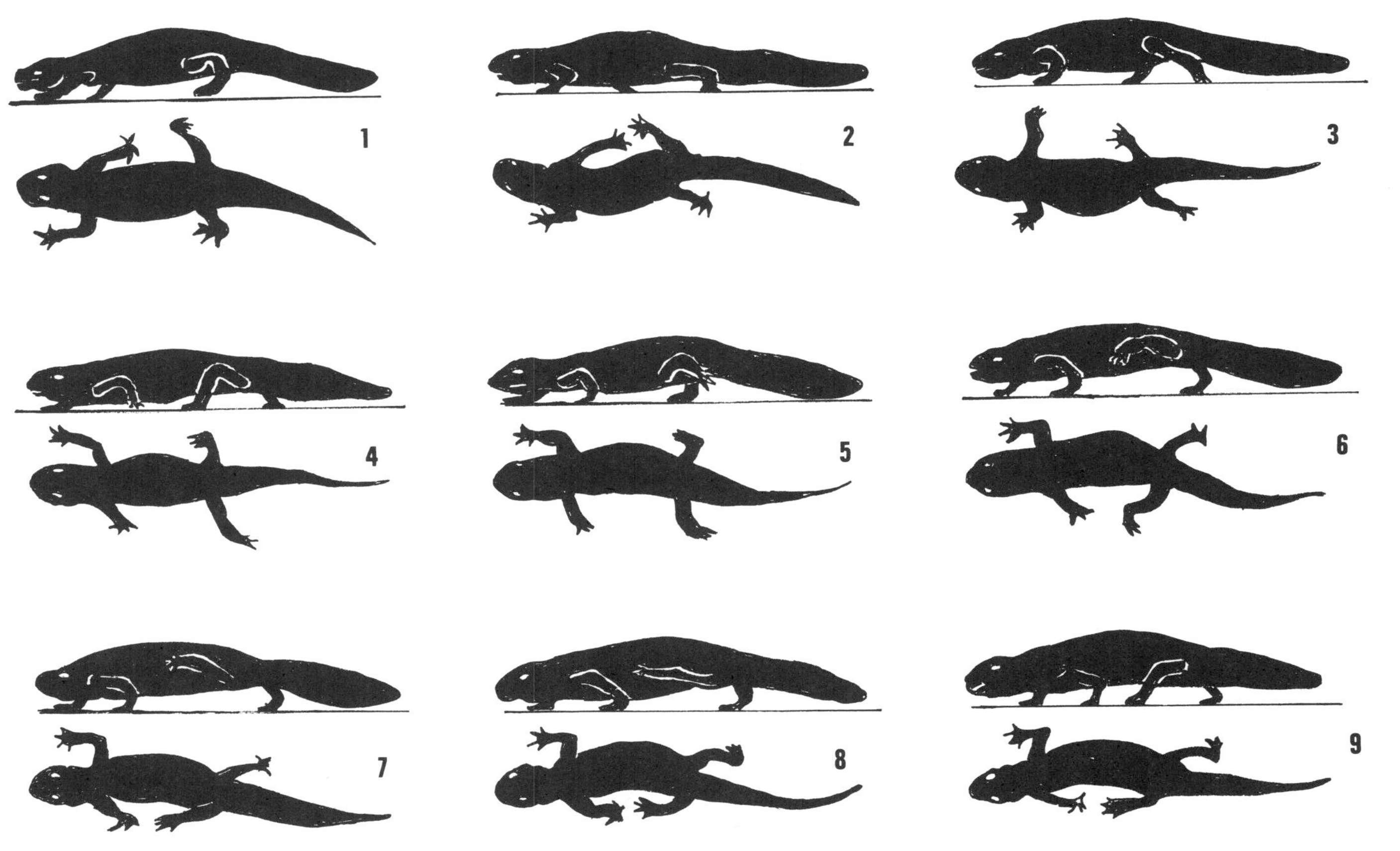


**Fig. 17.** Positions assumed by the limbs during locomotion of a newt. (Modified after Gray 1968)

Although about half of a crocodile's life is spent ashore, the animal lies sprawling and inert for most of the time, moving only in emergencies or in relation to the requirements of thermoregulation. When moving on land, crocodiles have three distinct gaits, described by Cott (1961) as the 'high walk', the 'belly run' and the 'gallop'. Of these, the first is the normal mode of locomotion, always used when the animal is hauling itself out of water or travelling unhurriedly overland. At such times it progresses, not like a lizard with the legs splayed out on either side, but with them swinging beneath the body and thus carrying the belly high off the ground. The sacrum is held higher than the shoulder, the head somewhat declined and the back is arched, only the lower surface of the tip of the tail trailing on the ground, while the legs move in diagonal pairs. The left fore and right hind limbs come forward together, followed by the right fore and left hind legs. The hind limbs are longer and stronger than the fore limbs and take most of the weight: the latter merely act as props to support the head and shoulders.

When disturbed on land, crocodiles race for the water in a crawling posture, sliding over the ground in the belly run, which is much faster than the high walk. Very occasionally, they may gallop, when the fore and hind legs work together in serial pairs. The body is carried forward by the thrust of the back legs while the fore limbs are extended to take the impact so that the crocodile bounds along 'like a squirrel' with a pitching motion of its body (Cott 1961).

During the course of their evolution, lizards have evolved fringes of elongated projecting scales on the toes, at least 26 times and in seven different families, according to Luke (1986; Fig. 18). The morphology of these fringes, which increase the surface area of the feet, varies acording to the type of substrate. Species that run on windblown sand usually have triangular, projectional and conical fringes, while riparian species that run on water tend to have rectangular fringes whose shape varies from wide to narrow. The association of the type of fringes with substrate has clearly evolved in many cases in response to selection pressures during locomotion, but differences in fringe morphology cannot be explained entirely on the basis of adaptation. When their shape is adaptive, however, fringes provide a good example of convergence since they have arisen independently so many times.

It is now well established that the mammals originated from the mammal-like reptiles, specifically the therapsid Cynodontia (e.g. *Cynognathus* Fig. 8). This evolutionary transformation invoked complex charges, not only in the skull but also in the limbs and their girdles. The latter permitted greatly

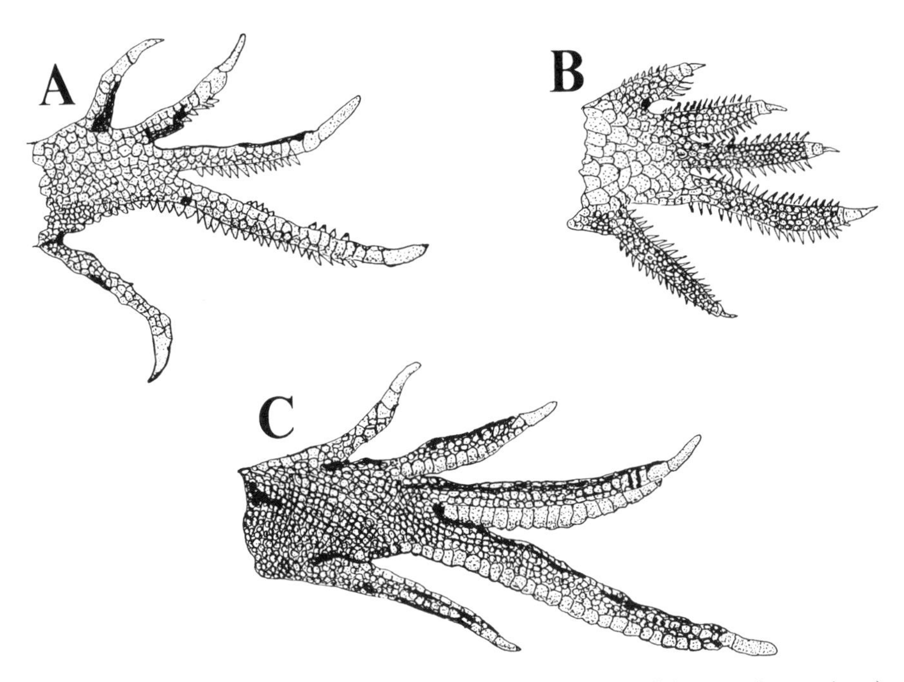


**Fig. 18.** Lizard toe fringes. **A** Triangular (*Uma notata*); **B** conical (*Teratoscincus scincus*); **C** rectangular (*Hydrosaurus pustulosus*). (Cloudsley-Thompson 1994; after Luke 1986)

increased freedom of movement of the legs and a change from a sprawling, lizard-like gait to a more erect type similar to the high walk of living crocodilians. The erect gait may have reduced stresses on the ankle and increased locomotory endurance, but this is by no means certain. Changes also took place in the feet and the number of phalanges was reduced to the mammalian formula of 2:3:3:3:3 for the five digits (Sues 1986).

Therapsid evolution was characterized by the following locomotory trends: (1) retention of a quadrupedal pose in all evolutionary lines; (2) adaptation of the limbs for locomotion, with the body raised above the ground; (3) expansion of the ilium and scapula; (4) retention of the five digits (Colbert 1986). Although some of the evolutionary trends of the Therapsida were conservative (e.g. retention of a quadrupedal stance and five digits), others showed considerable refinement, forecasting true mammalian locomotion which their far-off descendants were to inherit.

One of the principal reasons for the success of the dinosaurs was their ability to walk with erect limbs. Dinosaurs can be distinguished from other archosaur reptiles by the fact that their legs are articulated beneath the body so that not only are they able to support great weights, but also their stride is longer and more effective. According to Charig (1979), this posture can be traced back to large, semi-aquatic proterosuchians of the late Permian. These large, swamp-dwelling reptiles had, he believed, hind limbs modified as paddles which, in later archosaurs of the Triassic period, developed into powerful hind legs on land. Because of their superior locomotory ability on land, the early carnivorous dinosaurs of the Late Triassic became extremely efficient predators, and exterminated many herbivorous taxa, including the mammal-like Therapsida whose ecological niches were later occupied by dinosaurs. This argument, however, fails to account for the fact that the therapsids had themselves evolved a variable-gait walking technique in the late Permian. Furthermore, it is extremely unlikely that any carnivores should prey so heavily on other groups as to cause their extinction under natural conditions, when numbers of predators tend to fluctuate according to the abundance of their prey (Norman 1985). Modern reptiles are not heavy enough to require the weight of their bodies to be supported in this way.

## 4.2 Bipedal Locomotion

Urodeles do not seem ever to have perfected bipedal locomotion; but frogs and, to a lesser extent, toads have become specialized for jumping, as well as for swimming with their back legs. Anuran shoulder girdles can be divided into two broad categories, depending upon whether the cartilages connecting the coracoid bones are fused medially along their entire lengths (firmisternal condition) or along their anterior edges only (arciferal condition) (Fig. 19). The latter condition is more widespread and is found in 'toads' (Bufonidae, Pelobatidae, etc.) which tend to walk rather than to jump. The clavicles are well developed, the coracoids less so. The cartilage functions as a hinge, allowing a degree of independent movement between the two halves of the pectoral girdle (grossly exaggerated in Fig. 19) and freeing the arms. The firmisternal girdle of the frogs that hop and jump is a rigid structure which takes the shock of landing on the arms; and there is no independent movement of the two halves of the girdle. In the Apoda, there are no vestiges of limbs or of pectoral girdles. The subject has been discussed in greater detail by Noble (1931), Young (1981), Duellman and Trueb (1986) and others.

The characteristic jumping of a firmisternal anuran is an escape response that is employed typically in emergency situations. Unlike toads, most terrestrial frogs do not possess dermal poison glands. They need, therefore, to be

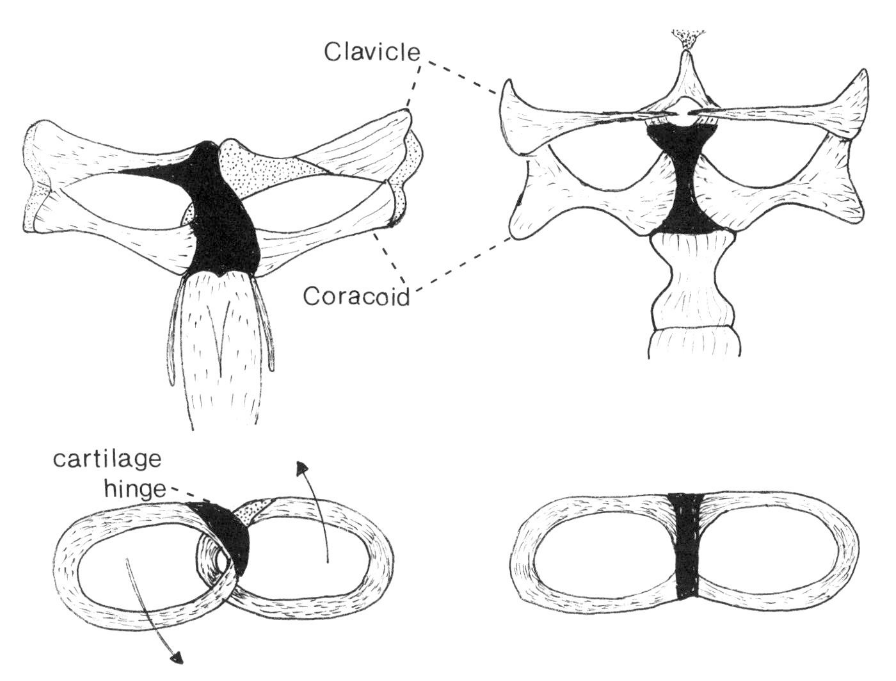


**Fig. 19.** Shoulder girdles of Anura seen from below. *Left* Arciferous condition of 'toads'; *right* firmisternal condition of frogs. Functional diagrams *below*. See text for further explanation

able to escape rapidly when threatened by predators. The energy to propel the body is provided entirely by the back legs: the fore limbs are held beside the body until the animal has reached maximum height. Both the height and length of a jump are extremely variable, but distances of over a metre can be covered by large frogs. In addition to the characters discussed in Chapter 1, long limbs and a lightly constructed posterior region to the body (Fig. 1) are also adaptations for jumping. When a frog takes off at an angle of 45° with the ground, about half the energy expanded is used in lifting the body against gravity, and the other half in providing forward movement. If the angle is less or greater than 45°, the length of the jump is somewhat reduced, but, for practical purposes, the velocity of take-off is much more important than the angle. Performance is also affected by the fact that the hind legs are considerably longer than the fore limbs. The frog does not become airborne until the centre of gravity of its body is higher above the ground than it is when the animal lands on its fore legs. The period of the upward phase of the jump is, consequently, shorter than that of the downward phase, and the total length of the jump is greater than it would be if the fore and hind limbs were the same length (Gray 1968). Thus we see that the modes of locomotion in Amphibia, as in other animals, are responsible for a diversity of structure.

The earliest reptiles were essentially quadrupedal animals, whose feet had five digits each – just as we have. As in modern reptiles, their limb girdles and legs were, however, more robust and powerful than those of amphibians. Salamanders move comparatively slowly on land, but lizards are able to run with agility and speed – as could some of the smaller dinosaurs, such as the bipedal coelurosaurs *Deinonychus* (Fig. 20) and *Velociraptor*, as well as the ornithomimosaurians *Ornithomimus*, *Gallimimus* and *Struthiomimus* spp. Whereas, in other carnivorous dinosaurs, the third toe was the longest, the second and fourth being considerably shorter, in *Deinonychus anterrhopus* the fourth and third toes were of equal length and the dinosaur walked and ran only on these two toes. The first toe was short, with a small backward-pointing claw; but the second toe bore a huge sickle-shaped claw from which the animal got its generic name which means 'terrible claw'. When the dinosaur walked, this claw was raised clear of the ground (Fig. 20). It could, however, have been swung through a very long arc and, when brought into play – in defence, or when disembowelling prey – it would have been driven backwards with the power of the entire limb. Not only would *Deinonychus* have had to balance on one leg, whilst doing this to a prey grasped by the long-

**Fig. 20.** Reconstruction of *Deinonychus* (length ca. 2.5–3.5 m), Early Cretaceous. (After Halstead and Halstead 1981)

clawed fingers of the hands, but also it might well have jumped and slashed at an enemy with the claws of both legs simultaneously. The discovery of the remains of about five individuals alongside those of a large herbivore weighing some six times as much as a single *D. anterrhopus* suggests that these dinosaurs may well have hunted in packs, just as lions do today, and were thus able to tackle prey very much larger than themselves (Halstead and Halstead 1981).

It seems probable that bipedalism in dinosaurs was presaged among the primitive archosaurs, including the ancestors of the crocodiles. This hypothesis is supported by the fact that some of the earliest members of the subclass, such as *Saltoposuchus* from the Upper Triassic, had fore limbs that were less than half the length of the hind legs. In the ancestral crocodile *Protosuchus*, this disparity was reduced to a ratio of about 1:1.5, and it is still less in modern crocodilians. Exceptional instances have been reported of living crocodiles running on their hind legs, and it is possible that *Pseudosuchus* was at times bipedal (see discussion of locomotion by Bellairs 1969).

Bipedalism in dinosaurs was by no means always correlated with speed, as seems to be the case among modern reptiles. The suborder Therapoda included all the bipedal carnivorous dinosaurs, some of which, such as the genera mentioned above, were speedy predators, while the carnosaurs were heavily built flesh-eaters. One of the latter, the renowned *Tyrannosaurus rex* (Fig. 21), was 12 m (40 ft) long and weighed about 6 tonnes. The horned ceratopsian dinosaurs represented a major group of ornithischians that evolved during the Upper Cretaceous. With few exceptions, they were heavily built quadrupeds, but the most primitive of them, the psittacosaurs or parrot-reptiles, were probably able to move on either two or four legs. Indeed, they represent one of the very few documented examples of a quadrupedal group of dinosaurs to have descended from a bipedal group (Halstead and Halstead, 1981).

Some modern lizards, such as the South American *Basiliscus basiliscus* (Fig. 22), the North American collared lizard (*Crotaphytus collaris*) (Iguanidae), the Australian *Chlamydosaurus kingii* (Fig. 22) and *Physignathus lesueurii* (Agamidae), are capable of running for considerable distances on two legs and are more than momentarily bipedal when travelling fast. Remarkably high speeds have been recorded – over 20 km/h in the case of the frilled lizard *C. kingii* – while the basilisk *B. basiliscus* has even been seen to dash for some distance across water. When at rest, however, the posture of the limbs and bodies of bipedal species is similar to those of any other lizards, and most of

**Fig. 21.** Reconstruction of *Tyrannosaurus rex* (length ca. 12 m), Upper Cretaceous. (Cloudsley-Thompson 1994)

**Fig. 22.** *Basiliscus basiliscus* (*above*) and *Chlamydosaurus kingii* (*below*) running bipedally. (Cloudsley-Thompson 1994)

the weight is taken by the ventral surface of the body. Nevertheless, the fore limbs play no part in tilting the body, which is lifted vertically by the extensor muscles of the hind legs without any preliminary running. Bipedal locomotion has occasionally been reported in the common iguana (*Iguana iguana*) as well as in the Australian perentie (*Varanus giganteus*). This is the largest lizard species in which it has been observed. Bipedal lizards fall into two main groups: those of open, rocky or sandy country and those such as the basilisks and iguanas, which live among trees and are both arboreal and semi-aquatic.

The bipedal locomotion of the basilisk and collared lizard has been analysed in detail by Snyder (1949) and summarized by Bellairs (1969). Although they may not become bipedal until they have reached high speed under natural conditions, these lizards are nevertheless able to assume a bipedal gait from rest or from a slow quadrupedal walk. The body is raised to an angle of up to about 45° and the fore legs are swung outwards. The action of the hind limbs is similar to that used in quadrupedal running, while flexion of the trunk helps to increase the length of the stride. The long tail is held off the ground and acts as a counterweight. The anatomy of lizards that can run bipedally differs only slightly from that of their relatives which always walk on four legs. There is a slight tendency only for the length of the trunk to be reduced and for the disparity in the total length of the fore and hind limbs to be increased (Snyder 1962).

## 4.3 Adaptations for Running on Sand

Lizards that inhabit sandy deserts show three principal types of adaptation: they are either sand swimmers and burrowers (see below) or sand runners, which, as we have seen, have fringes of elongated scales on their toes (Fig. 18). Several different genera have become adapted in this way. These include *Uma* (Iguanidae), *Phrynocephalus* (Agamidae) and *Stenodactylus* (Gekkonidae) (Fig. 23). Gekkonid lizards are an important element of the desert fauna, but they are somewhat rare among dunes. Nevertheless, modifications of the feet for sand running are found in the gecko subfamilies Diplodactylinae and Gekkoninae. These provide excellent examples of convergence (Luke 1986).

Three types of pedal adaptation are associated with life among dunes: fringed toes, spinose, swollen plantar surfaces and webbed feet. The first is associated with the occupation of loose, leeward dune slipfaces, the second

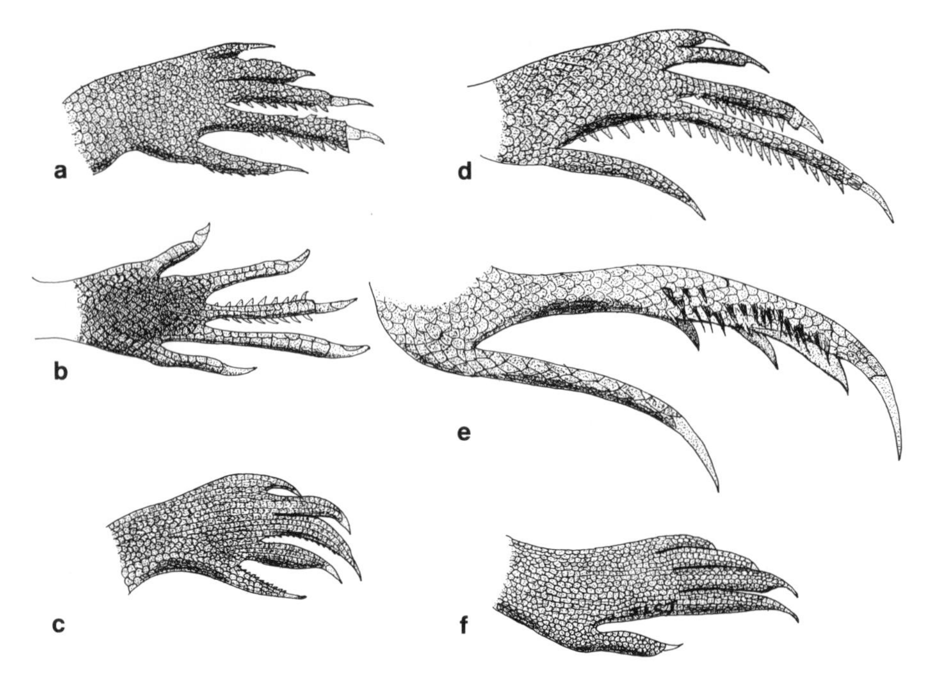


**Fig. 23.** Feet of desert lizards showing scales. **a**–**c** Fore feet; **d**–**f** hind feet. **a**, **d** *Phrynocephalus* sp.; **b**, **e** *Uma* sp.; **c**, **f** *Stenodactylus* sp. (Cloudsley-Thompson 1991; after P. A. Buxton 1923)

with sandy substrates in general, and the third with burrowing in the compacted sand of windward dune faces. Although webs and fringes increase the surface area of the feet and aid in locomotion on sand, both appear primarily to facilitate movement below the sand surface, and burrowing. For instance, the webbed feet of the African *Palmatogecko rangei* are used mainly for digging (Bauer and Russell 1991).

## 4.4 Adaptations for Burrowing

Certain taxa, both of amphibians and reptiles, have, evolutionarily speaking, lost their legs and move by serpentine motion. Caecilians (Gymnophiona or Apoda) comprise an order of very specialized amphibians of which only about 80 species now exist. They are worm-like inhabitants of tropical regions, where they burrow in the soil. Their major adaptations for this mode of life consist in the disappearance of the legs and elongation of the body. This is segmented by primary annular grooves. Some species have secondary

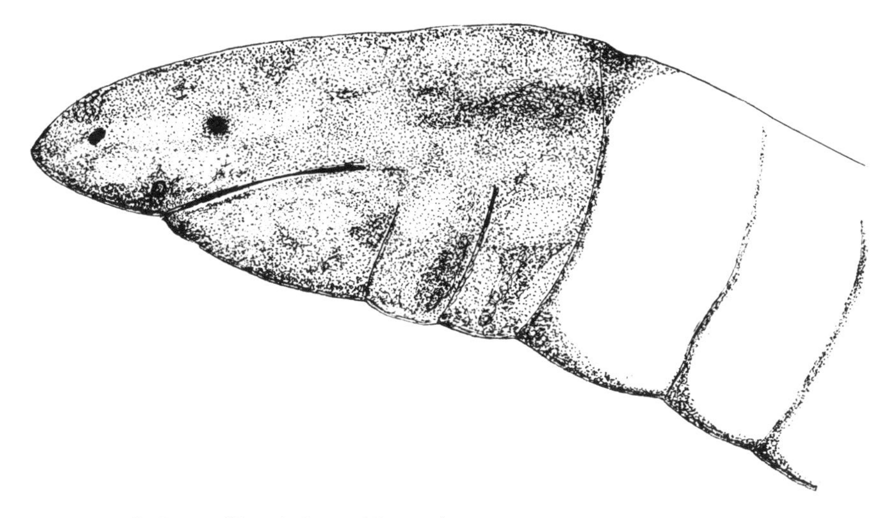

**Fig. 24.** Head of a caecilian. (After Noble 1931)

grooves dividing the primary annuli, but none have tertiary grooves. In the more primitive genera, the grooves contain scales. These are almost certainly an inheritance from their Carboniferous amphibian ancestors, but no Palaeozoic or Mesozoic fossils are known. The rear end of the body is short, pointed and capped with a terminal shield. The eyes are small, covered with skin or hidden beneath the bones of the skull. Protrusible olfactory tentacles, found on the sides of the face below and between the nostrils and the eyes, are characteristic of all species (Fig. 24). The function of these is the detection of prey by smell.

Several unrelated species of lizard have become adapted for burrowing in soft earth or sand. In some instances, this is facilitated by the presence of specialized scales on the feet, but, in the majority of cases, the animals propel themselves through the soil by serpentine movements of the body. The limbs of burrowing lizards are usually small, and the body long and thin. In the European slow-worm (*Anguis fragilis*) and the much larger glass lizard (*Ophisaurus apodus*) (Anguidae) the legs are entirely absent. One slow-worm of Oligocene times had a skull 12.5 cm long and is appropriately named *Glyptosaurus giganteus*. Fossils of Anguidae are known from the Cretaceous period, indicating that the loss of limbs has evolutionary advantages. Legs are also absent from snakes. Indeed, there seems to be little doubt that the ancestors of snakes lost their limbs as an adaptation to burrowing – a mode of life that the majority have now forsaken.

The genus *Ophisaurus* has an almost worldwide distribution, indicating that it has been in existence for at least 30 my, dating back to Oligocene times. The name 'glass lizard' refers to a myth that the animal breaks into pieces like a glass rod when struck, but can later reassemble itself unharmed from its experience. This idea is probably based on the fact that, although their tails are much longer than the body, they can be autotomized just like those of other lizards – a subject to which we shall return later (Pope 1955).

In the Palaearctic three-toed skink *Sphenops* (= *Chalcides*) *chalcides*, the body is extremely elongated and snake-like, but there are vestigial limbs, so small as to be scarcely visible. The legs are somewhat larger in some other North African and Middle Eastern members of the genus *Sphenops*, but, even so, are usually too small to be of much use in locomotion. Legs are absent from the Greek legless skink *Ophiomorus punctatissimus*.

About 160 species of Amphisbaenidae or worm lizards have been described. They get their name from the Greek works *amphis* which means 'on or at both sides' and *baina*, 'one step'. This 'going at both sides' is supposed to mean 'going at both ends' and, in their tunnels, amphisbaenians do indeed move both forwards and backwards. Whereas most burrowing reptiles move, like snakes, by serpentine movements, amphisbaenians are unusual in that their mode of progression is essentially similar to that of earthworms. The body, too, has an annulated appearance. Waves of muscular contractions pass along the body, which is locked to the ground where the underlying muscles are shortened and the body thus expanded. Those regions of the body which are actively shortening pull the posterior portions forward and then themselves are pushed forward as the soil is gripped behind them. This type of movement is fundamentally similar to the rectilinear movement of certain snakes described below, but, whereas in a boa or viper the waves of contraction are restricted to the ventral musculature, in amphisbaenians they are also exhibited by the dorsal muscles. This is an advantage to a burrowing animal whose entire surface is in contact with the surrounding soil (Gray 1968; Bellairs 1969).

The locomotion above ground of legless amphibians and lizards is similar to that of snakes, which will be discussed in the following section. Whereas these, with the exception of Amphisbaenidae, burrow into the ground with serpentine movements, burrowing frogs and North American spadefoot toads (*Scaphiopus* spp.) dig with their feet. Each year, these toads emerge on the first night of heavy rain to spawn in the newly formed pools. Spadefoot toads are nocturnal, and spend the day buried a few centimetres below-ground, emerging to forage on nights following rainfall when the desert surface is relatively damp. In September, they retreat into deep hibernation burrows where they remain inactive for the next 9 or 10 months (Tinsley 1990).

The ability to dig into sand is often critical for reptiles that inhabit dunes, both as a means of sheltering from the sun and of escaping from enemies. Sand swimmers of desert regions include *Sphenops* spp. in North Africa and the Middle East, iguanid lizards of the genus *Uma* and shovel-nosed colubrid snakes (*Chionactis* spp.) in the Mojave desert, as well as the lacertid lizard *Aporosaura anchietae* and the sand viper *Bitis peringueyi* in the Namib. (Many of the lizards have reduced or absent limbs.) These reptiles either use low amplitude, high frequency lateral undulations to sink into the sand, or dive into it, head first. They usually possess morphological and physiological mechanisms for breathing under the sand, because they do not construct permanent burrows and are buried in the substrate. The nose or rostrum is pointed and shovel-shaped. The nostrils may be directed upwards, instead of forwards, as a protection against the entry of sand, while valve-like closure of the eyes, nostrils and mouth enables them to dive into loose sand as though it were water. Their bodies are covered with smooth scales which engender little friction, and movement is accomplished by serpentine wriggling, as already mentioned.

Infiltration of sand into the space left by exhalation is a major problem for sand swimming reptiles. Different species have evolved different morphological features in response to it, but the major adaptation consists of shielding the breathing movements from the surrounding sand. Thus, when *Uma notata* (Iguanidae) buries itself, it extends its fore legs backwards against the sides of the body so that they form a rigid roof which excludes sand, and beneath which the lizard can breathe. The horned lizard *Phrynosoma platyrhinos*, another iguanid of North America, does the same, but the ribcage is released, producing folds of skin along the posterior part of the trunk which provide additional shielding (Pough 1969). Breathing by vertical movements of the venter is a common feature of sand-dwelling reptiles. This is illustrated by the fact that the Saharan *Sphenops sepsoides* and *Scincus officinalis* have concave ventral surfaces, whereas *S. ocellatus*, which inhabits the firmer sand of the oases, has a rounded body (Mosauer 1932). The North American legless *Anniella pulchra* (Anguidae) constructs open burrows in firm soil, but ventures freely into dry sand (Miller 1944): it has a rounded body with a moderate ventral cavity. On the other hand, the African skink *Lygosoma* (= *Riopa*) *sundevelli* retains the basic breathing pattern of lateral expansion of the body, but this is screened by ridges which appear upon exhalation. The lower ridge results from the inflexibility of the ventral scales, while the stiffness of the dorsal scales and ribs contribute to the formation of the upper ridge. In consequence, respiratory movements are restricted to a narrow depression (Pough 1969).

In contrast, burrowers have relatively shorter bodies, produce weak undulations, but show modifications of the feet for digging. Many, including most dune-dwelling geckos, excavate discrete burrows and consequently do not require modifications for breathing under the surface of the sand. The burrowing North American colubrid sand snakes *Chilomeniscus cinctus* and *Chionactis occipitalis* both have concave ventral surfaces, but, whereas the Indian sand boa *Eryx conicus* shows few morphological adaptations for burrowing, in the larger *E. johni* the head is pointed, with a countersunk lower jaw and a pronounced rostrum. *E. johni* breathes by vertical movements of the venter when it is buried, but *E. conicus* does not (Pough 1969).

The different shapes of the heads of amphisbaenids are associated with their burrowing techniques. For instance, members of the genera *Amphisbaena*, *Blanus*, *Cadea* and *Zygaspis* burrow by ramming the soil with their heads. *Leposternon* and *Monopeltis* spp. tip their heads downward, thrust forward, and then lift their heads, while *Anops* and *Ancylocranium* spp. have heads that are compressed laterally. Individuals of these genera alternately ram their heads forward and swing them from side to side (Zug 1993). The body is then moved forward in the manner already described.

The eyes of many reptiles, especially burrowing forms, are shielded by transparent 'spectacles'. The lower eyelid is modified to form a window: in some species there is a single, circular, transparent disk, while in others the window occupies the whole eyelid which is permanently closed and fused with the upper lid. In addition to their obvious value in excluding foreign particles from the eyes, spectacles are an important adaptation to water economy because, in reptiles that lack them, evaporative loss from the moist eyes contributes considerably to total transpiration (see Chap. 11 and the discussion of burrowing adaptations among desert lizards in Cloudsley-Thompson 1991).

## 4.5 Limbless Locomotion on Land

Although the disappearance of legs from some amphibians and reptiles is primarily an adaptation for burrowing, many legless forms have secondarily returned to surface dwellings. Although some are aquatic, nearly all the caecilians have retained their burrowing habits. This may be correlated to some extent with water conservation. Most snakes and legless lizards, on the other hand, now live aboveground. It is with these – and specifically with snakes which best illustrate the principles involved – that the present section is concerned.

Snakes exhibit four main types of terrestrial locomotion: (1) serpentine or undulatory motion; (2) concertina movement, in which friction is used; (3) rectilinear locomotion, as in pythons and boa constrictors, and (4) sidewinding by which speed can be achieved with little friction (Gans 1962). Under appropriate conditions most snakes can make use of all these except for rectilinear locomotion, but they will be discussed separately in the following paragraphs.

Serpentine movement is used by nearly all snakes and legless lizards when they glide through grass, between stones or across sand. The animals move by throwing their bodies into a series of lateral or horizontal undulatory waves which flow continuously from head to tail. These are produced by a flow of muscular contractions and relaxations which pass down the body. The muscles on one side of a particular region of the body contract while those on the other side relax. The process has been analysed by J. Gray and H. W. Lissman (Gray 1968) who showed that the propulsive force which moves a snake is derived from the thrust made by the inner side of each curve of the body against projections from the substrate – stones and other objects, the stems of plants and irregularities in the soil surface. The dependence of the snake upon these projections is illustrated by its inability to crawl efficiently across polished surfaces. It can, however, glide past a series of smooth pegs projecting from the surface of a smooth board (Fig. 25). (In contrast, when they are swimming, the bodies of marine ragworms such as *Nereis* spp. exert maximum pressure with the parapodia that lie on the *outside* of the curves. Parapodia are paired lateral projections extending from the body segments. For this reason, undulations begin from the tail end in ragworms and from the head in eels and snakes.) The sinusoidal curves of a snake's body when it is gliding by serpentine movement are stationary relative to the ground: it is the snake's body that moves, each part closely following the path of the part in front (see Gans 1962; Bellairs and Carrington, 1966; Gray 1968; Bellairs 1969; Parker 1977; Cloudsley-Thompson 1978a).

When a snake crawls through a relatively narrow tunnel, such as a drainpipe, which is too narrow to permit undulatory locomotion it adopts concertina movement. The body is bunched up and thrown into a number of

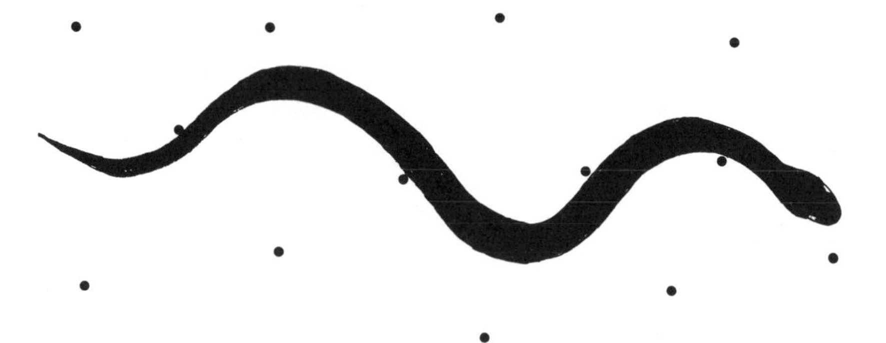

**Fig. 25.** Snake gliding past a series of smooth pegs projecting from the surface of a smooth board

small curves which are pressed against the walls of the tunnel. Where the tunnel narrows, the number of curves in the body of the snake increases and their size decreases. The coils that are pressed against the wall of the tunnel anchor that part of the body while the snake straightens out in front. The process is continually repeated so that the reptile moves steadily forwards. The same method is employed when a snake is moving over a flat surface and the weight of the coiled part of the body is sufficient to give it purchase.

Another type of movement is practised by pythons, boas, large vipers and other thick-bodied snakes. It is termed rectilinear movement and does not involve any flexion of the body. The snake is able to creep slowly forward like an amphisbaenian with its body almost straight. This is possible because the skin and ventral scales are relatively loose and freed from the underlying tissues. Successive groups of scales are raised, drawn forward, and then placed on the substrate. The rest of the body is then hitched forward over them by waves of muscular contractions which pass down the ventral abdominal wall. These contractions are timed so that the snake glides forward at a steady speed although the groups of scales move in jerks. Compared with serpentine movement, rectilinear movement is relatively slow. It is employed when a snake is stalking its prey or moving over flat ground.

Finally, we come to sidewinding or crotaline movement, characteristic of viperid snakes that inhabit sandy deserts. As in serpentine movement, the snake propels itself by propagating waves of curvature posteriorly along its body. Instead of moving over the ground with the axis of progression coinciding with the axis of the waves, however, the body moves sideways at a

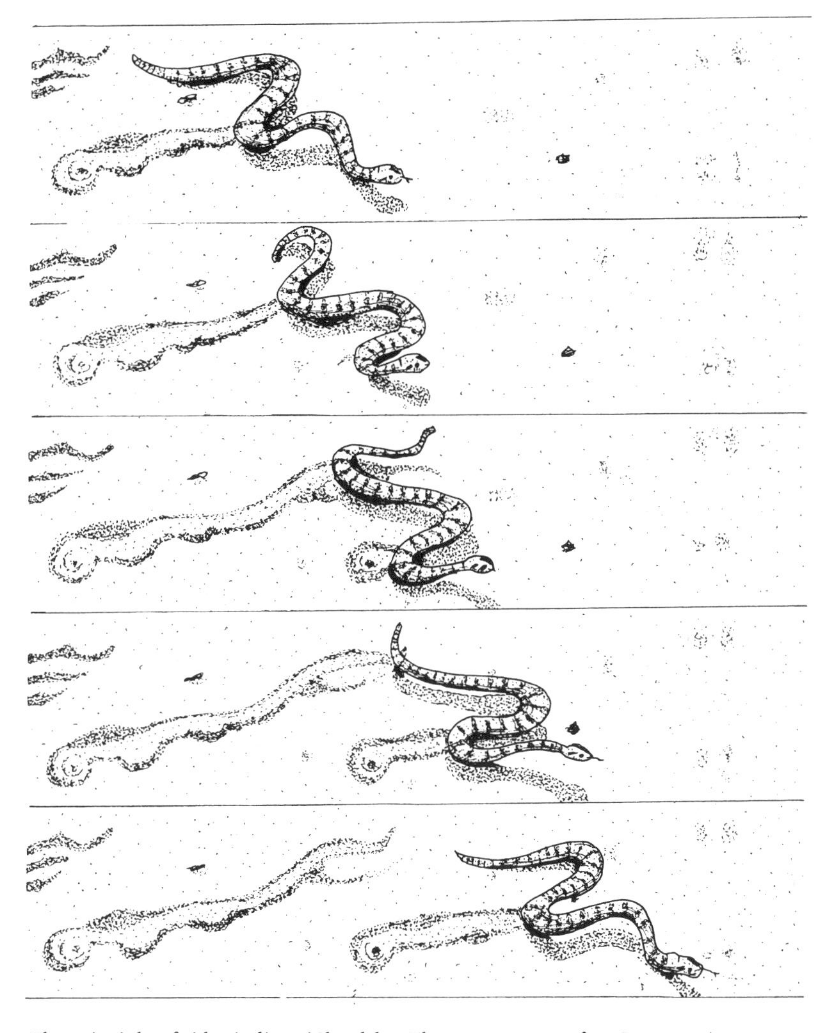

**Fig. 26.** The principle of sidewinding. (Cloudsley-Thompson 1994; after Gans 1970)

considerable angle to the axis of the waves. A sidewinding snake achieves firm static contact by moving so that its body lies almost at right angles to the direction in which it is travelling, and its tracks in the sand appear as a series of parallel lines each at an angle of about 60° to the snake's direction of movement (Fig. 26). Static friction is thereby employed without sacrificing speed. Not only does sidewinding allow a snake to cross hot desert without overheating because so much of the body is raised above the ground, but also it is quicker than other forms of locomotion on a loose substrate. Furthermore, it gives a misleading impression of the direction in which the snake is moving so that it is able to surprise its prey by the speed with which it approaches.

The locomotion of amphibians and reptiles has been reviewed by many of the authors cited above, but, of these, the most detailed accounts are those of Bellairs (1969) and Gans (1970) on reptiles, and of Gray (1968) who analysed the locomotion of animals of all kinds.

# 5 Arboreal, Aerial and Aquatic Locomotion

Escape from enemies and capture of prey are the two components of natural selection that have most influenced the morphology and lifestyles of amphibians and reptiles. The earlier terrestrial vertebrates must have included forms adapted to escape from predators and obtain food by climbing rocks and, in due course, vegetation. Today, even fishes such as mudskippers (*Periophthalmus* spp.) have scansorial adaptations and are able to climb up the aerial roots of mangrove trees. Arboreal life is of more than passing interest, for it is probable that most flying vertebrates, except for flying fishes, were and are derived from scansorial ancestors.

## 5.1 Scansorial Adaptations

Arboreal amphibians include some of the stegocephalians of the Coal Measures as well as several families of modern tree frogs. Not many salamanders are arboreal - possibly because they are more prone to desiccation than are frogs - yet, surprisingly, the only North American tree-dwelling salamander is a member of the lungless family Plethodontidae. This is the oak salamander (*Aneides lugubris*), which reaches a length of 15 cm or more. The genus *Oedipus*, found in the neotropics, also contains a number of other arboreal species. The adaptations for climbing found among arboreal salamanders include circular toe pads (Fig. 27a), although these are not confined to tree-dwelling species (Smyth 1962). In *Aneides* spp. the terminal phalanges are Y-shaped and bent downwards, while in *Oedipus* spp. they are joined by a thick pad.

Toe pads are also characteristic of arboreal frogs (Fig. 27b), but some, such as the African *Chiromantis xerompelina* and various species of the South American genus *Phyllomedusa*, have reduced webs, while both hands and feet are transformed into organs that grip the branches upon which the frogs sit.

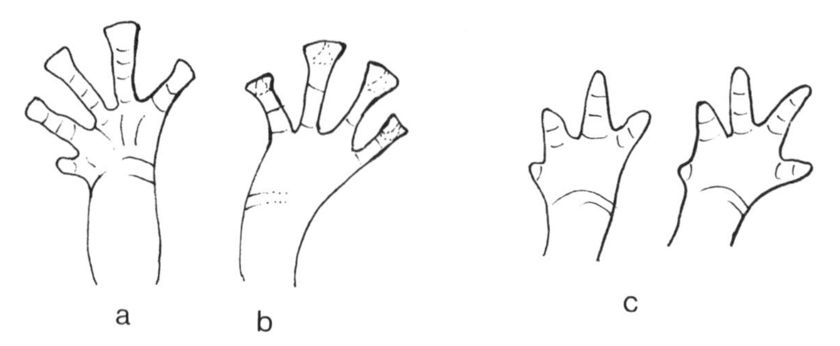


**Fig. 27.** **a** Elongated toes with squarish tips and circular pads of a climbing salamander (*Aneides* sp.); **b** toe pads of a hylid frog, an adaptation for clinging and climbing; compared with **c** typical ampibian toes – four on the front feet, five on the hind feet – of a terrestrial salamander (*Plethodon* sp.). (After Stebbins and Cohen 1995)

Grasping is aided by the fact that these species, unlike most other Salientia, have opposable 'thumbs'. Other tree frogs have some webbing between their toes and usually between their fingers as well, depending upon the species. Arboreal amphibians, in general, tend to have rather long legs and flattened bodies, both characters being of adaptive value. In addition, colour change is particularly well marked amongst them.

Although many different families of frogs are tree-dwellers in other parts of the world, North American arboreal frogs are all confined to the Hylidae. Not all the species of this family are tree climbers, however: many live on the ground or on low plants and shrubs. In general, the larger tree frogs tend to climb higher than smaller species, but there are some exceptions. The green tree frog (*Hyla cinerea*), for instance, does not climb so high as either the squirrel tree frog (*H. squirella*) or the pinewoods tree frog (*H. femoralis*), yet the last two species are both smaller than *H. cinerea*, although all three are larger than the spring peeper (*H. crucifer*) and other non-climbing hylids (Smyth 1962). This is surprising because small, non-climbing species could probably climb more easily than the larger species – including the largest of all, *H. vasta*, which climbs high but inefficiently. Possibly the smaller surface to volume ratio of the larger species may render them less vulnerable to desiccation at higher altitudes where the saturation deficiency of the atmosphere is greater.

No such problem faces scansorial reptiles, which have much lower rates of transpiration than amphibians. Arboreal adaptations are very common among the Squamata of tropical rainforest. They include prehensile tails. These are characteristic of chameleons and arboreal vipers, such as the oriental green pit viper (*Trimeresurus albolabris*) (Fig. 28).

**Fig. 28.** Green pit viper (*Trimeresaurus albolabris*) represented upon a dead branch in order to show clearly a characteristic attitude with the tail grasping the branch and the head prepared to strike (length ca. 50 cm). (Cloudsley-Thompson 1994; after Borradaile 1923)

The twig snake (*Thelotornis kirtlandii*) (Colubridae) (Fig. 29) of central and southern Africa has a long, whip-like body which it can stiffen, by tightening all its muscles, in order to bridge the gap between one branch and another. The common name, twig snake, refers to the habit of maintaining the body rigid, extended from a branch and swaying to and fro. This behaviour, coupled with its pale brown and greyish body streaked and spotted with white, dark brown, grey and black, which resemble the colour of bark, renders the snake extremely inconspicuous in its natural environment. The generic name, derived from the Greek, alludes to its mythical ability to charm the birds which form part of its prey. Like its back-fanged relative the boomslang (*Dispholidus typus*), the twig snake is known to have caused the death of several human beings. The ventral scales of tree-dwelling snakes are usually stiffened by transverse keels which provide added traction on rough bark.

The tails of most chameleons, like those of arboreal snakes, are modified for prehension. Chameleons are most unusual, however, in that their tail coils vertically like the spring of a clockwork motor. The scales on the lower surface of a chameleon's tail are spiny and give a good grip on branches, while the upper surface, in contrast, is quite smooth. The longitudinal muscles of the tail give off a series of tendons at regular intervals. These run backwards for

**Fig. 29.** Twig snake (*Thelotornis kirtlandii*) (length ca. 1 m). (Cloudsley-Thompson 1994; after Gadow 1901)

a short distance before they become attached to more posterior caudal vertebrae. These are, themselves, shaped in such a way that they can coil in a vertical plane. Thus, when the tail muscles contract, all the tendons are pulled simultaneously, and the movement of coiling or uncoiling is distributed evenly through the entire length of the tail. The limbs of a chameleon are comparatively long, and the shoulder and hip joints move freely. The ligaments of the shoulders are more lax than in other lizards. Consequently, the arm and wrist are so mobile that the hand can grasp objects in any direction relative to the animal's trunk which, itself, always remains upright. The digits of the hands and feet, like those of some arboreal frogs, are opposable so that they can grasp twigs. This character, too, is unique among lizards to chameleons. In the hand there are two digits on the outer side and three on the inner, whereas in the foot there are three digits on the outer and two on the inner side (see description in Bellairs 1969).

The adaptations of arboreal geckos for climbing are very different from those of chameleons. Geckos are able to cling both by means of the sharp, recurved claws on their toes, which are present in most species, and by the

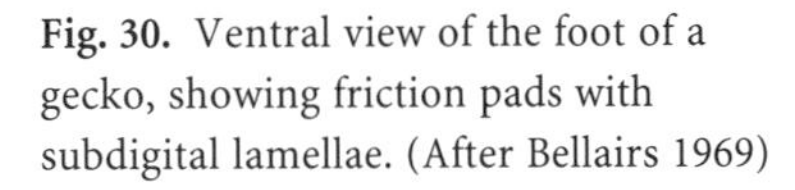

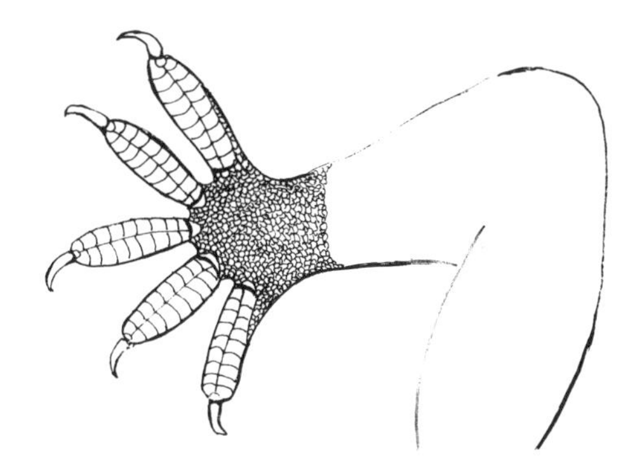

**Fig. 30.** Ventral view of the foot of a gecko, showing friction pads with subdigital lamellae. (After Bellairs 1969)

action of special scaly expansions situated proximally to the claws on most or all of the digits. The claws are compressed laterally and, without them, a gecko cannot adhere to the underside of horizontal surfaces and has difficulty in climbing rough vertical surfaces such as rocks and brick walls. It can still stick to smooth vertical surfaces such as glass, however, through the action of the digital expansions. These expansions or pads are covered on their undersurfaces by a row or rows of wide scales known as 'lamellae'. Each lamella, in turn, is covered, or partly covered by large numbers of setae – fine, hair-like structures (Fig. 30). In most species of gecko, the lamellae are arranged either in a single row or in two rows, overlapping each other at the edges like the ventral scales of a snake. Their precise arrangement is of systematic importance. In *Sphaerodactylus cinereus*, the adhesive pad is restricted to a single terminal scale and bears approximately 6000 setae, whose ends are shaped like inverted cones. These provide the friction that enables the gecko to walk on vertical glass plates (Röll 1995). Even when a gecko is hanging upside down, it is able to exert a horizontal force by the pinching movement of its digits. As it is held by friction, rather than by suction or adhesion, a large number of setae are more efficient than a smaller number of larger setae would be. Furthermore, suction requires special muscles to lift parts of the sole off the surface and adhesion requires a fluid, neither of which is present in the pads of a gecko. Apart from in the Gekkonidae, setae have also been found in arboreal Iguanidae, Scincidae and Chamaeleonidae, but they are shorter and do not branch (Bellairs 1969).

The arboreal habit is sometimes facultative, sometimes obligatory, among snakes. Remarkable convergences in morphology, physiology and behaviour are to be seen between quite unrelated arboreal taxa. The arboreal habitat presents various challenges, such as discontinuous and potentially unstable substrates, increased risk of predation, rigorous microclimatic conditions in some plant communities, and disturbances to the circulation of the blood due to the gravitational effects of vertical postures. Evolutionary adaptations to these challenges include reduction in mass, increase in body length, marked cryptic and usually green coloration, locomotory specializations, great visual acuity, and modifications of the cardiovascular system (Lillywhite and Henderson 1993).

## 5.2 Jumping

The ability to jump is of adaptive value to many scansorial and aerial animals, most of which take to the air with a leap. Not only can frogs and bipedal reptiles jump, but also some four-legged cursorial types and, perhaps unexpectedly, even snakes. Jumping is particularly well developed in snake lizards (family Pygopodidae) of the genera *Aclys*, *Delm*a and *Ophidiocephalus*: *D. tincta* is one of the most excitable species known. When an individual is disturbed, it is said to twist its body and jump about in a fantastic display of aerobatics.

Over 30 species of snake lizards are distributed throughout Australia, Tasmania and New Guinea. Their tails are extremely long, exceeding the length of the head and body, and there are no front limbs. The back legs are reduced to mere flaps which are often quite small, and their movement is snake-like. The Pygopodidae is the only reptile family endemic to the Australasian region: and it is allied to the geckos. Although saltation may account for a high rate of success in escaping predation, locomotion is probably not the sole function of the response. Rather, it appears to be a mechanism for startling or disorienting potential predators, and for eliciting misdirected strikes. Furthermore, the long tail is readily autotomized. When first broken off, its lively movements serve to confuse the enemy long enough for a snake-like lizard to make its escape. Other lizards, such as the Asian desert *Phrynocephalus mystaceus* (Agamidae) are also capable of jumping.

It is by no means unexpected that bipedal, and even quadrupedal, amphibians and reptiles should be able to jump. It is more surprising to learn that some snakes are able to do the same. For instance, the 'jumping viper' (*Bothrops nummifer*) (Crotalinae), which receives its Spanish name 'mano de piedra' from its resemblance in appearance to an implement used in crushing corn for tortillas, is particularly savage. A stocky, terrestrial pit-viper of Mexico, Guatemala, Honduras, San Salvador and Panama, it spends the day basking on or sheltering beneath or inside fallen logs, piles of leaves and so on. At dusk, it becomes more active and may forage actively or wait in ambush for the small rodents, lizards and frogs on which it preys. It lives up to its name by sometimes striking with such force that it actually leaves the ground and makes a short jump. When striking from the side of a log, or from a bank, like *Bitis* spp. it may even be able to propel itself for more than a metre. It bites defensively as well as in offence and, when thoroughly frightened, flails about, striking wildly, and turning quickly to keep the enemy in view. Similar behaviour has been observed in other ground-dwelling species of the genus, and provides an explanation for their popular name, 'lance snakes'. Some speedy Colubridae, such as the large green rat snake (*Pityas nigromarginatus*) of southern Asia, can hurtle through the air down steep slopes for considerable distances in pursuit of prey or escape from enemies. *Chrysopelea* spp. (see below) are also able to jump from the branches of the trees among which they live. Finally, cryptozoologists have reported a mysterious jumping snake from the environs of Sarajevo. Could this be *Coluber viridiflavus*, the western whip snake? Many scansorial amphibians and reptiles begin their glides with a jump.

## 5.3 Gliding and Flying

Gliding among amphibians was first reported during the nineteenth century by A. R. Wallace in the Malaysian frog *Rhacophorus nigropalmatus* to a somewhat sceptical world. Cott (1926) was the first to study the flight of an aerial frog. He showed that the Brazilian *Hyla venulosa* fell slowly, always belly down, at a gradient of about 60°, with its legs spread laterally. The European *H. arborea* and *Rana temporaria*, although about the same size and no worsely equipped structurally to check their fall, fell vertically, gyrating wildly and landing heavily. Thus, the first requirement for parachuting and gliding seems to be behavioural, through the development of attitude control, rather than morphological.

No modern amphibians or reptiles are able to fly, but, as we have seen, some can parachute and glide for considerable distances (Fig. 16). Of these, the most remarkable are the flying dragons (*Draco* spp.: Varanidae) of South-east Asia, flying geckos (*Ptychozoon* spp.: Gekkonidae) of the same region,

**Fig. 31.** Flying dragons (*Draco volans*) (length ca. 15 cm). (Cloudsley-Thompson 1994; after Gadow 1901)

and flying snakes. In *D. volans* and related species, the body is depressed and can be extended sideways by five or six elongated ribs. These flying lizards glide through the air, buoyed up by scaly fringes which run along the sides of the head, limbs, body and tail, and by webs between the digits (Fig. 31). Their flight is skilfully controlled, and a courting pair has even been observed to perform sexually during a full aerial roll (Savile 1962). *Kuehneosaurus* sp. (Fig. 32) of the Upper Triassic period, about 220 my ago, was also a well-adapted flying lizard.

**Fig. 32.** Reconstruction of *Kuehneosaurus* sp., Upper Triassic (length ca. 75 cm). (Cloudsley-Thompson 1994; after Charig 1979)

Fossils of another Triassic reptile, *Sharovipteryx mirabilis*, show that this, too, was a glider. Its hind limbs were more than three times as long as the fore legs. Their proportions exceeded those of agamid lizards, such as *Otocrypsis* and *Amphibolurus* spp., which run and jump with the trunk elevated so that the fore limbs may, or may not, touch the ground between strides. *S. mirabilis* was probably a runner and jumper too which, like modern flying lizards, inhabited the terminal branches of trees. The membrane or 'patagium' between its limbs and elongated tail almost certainly served as a gliding membrane. It may also have aided camouflage, reducing the outline of the creature, as, for instance, do the lateral flaps of certain Pacific geckos (Cloudsley-Thompson 1994).

Back-fanged colubrid tree snakes of the genus *Chrysopelea* (e.g. *C. ornata*) not only scale the trunks of forest trees by pressing the coils of their bodies against irregularities in the bark, but also can glide obliquely through the air with bodies rigid and ventral surfaces concave so that they present maximum resistance to the air. They can also spring from one branch to another by coiling their bodies and then rapidly straightening them.

Only four taxa of animals have acquired powered flight with true wings. These are insects, pterosaurs, birds and bats. Whereas the early pterosaurs would have been gliders, later forms possessed large, keeled breastbones for the attachment of the powerful wing muscles necessary for true flapping

**Fig. 33.** Long-tailed pterosaur *Rhamphorhynchus* sp., Jurassic (length ca. 60 cm), showing attachment of the wing membrane

flight. The pterosaurs (Fig. 33) first appeared in the fossil record during the Upper Triassic, and flourished until the close of the Cretaceous period, some 155 my later. Many of them were no larger than pigeons, but some of the Cretaceous pterodactyls were huge. *Quetzalcoatlus*, appropriately discovered in Texas, had an estimated wing span of 11 m and was the largest flying animal known. Computer studies of *Pteranodon* have shown that pterosaurs were adapted for low-speed flying and, like large birds, would have soared on weak thermals. Their low stalling speeds enabled them to take off by facing into the wind with wings extended, and they probably did not fly when strong winds were blowing. The long crest on the back of the head (Fig. 34) would have counteracted the twisting effect of wind on the beak (McGowan 1991).

The wings of the pterosaurs were not attached to the feet, as are those of bats, but probably to the pelvis with the membrane supported by a single fourth finger. This gave the animals a more attenuated shape, and the result-

**Fig. 34.** Reconstruction of *Pteranodon* sp., Upper Cretaceous (wing span ca. 7 m). (Cloudsley-Thompson 1974)

ing wing loading resembled that of modern birds. Except for the largest species, they were active fliers and, unlike bats, walked bipedally. Some species may have been scavengers like vultures, but it seems likely that many would have swooped down from the sky to catch their prey whilst in flight. Others might have picked up small animals from the ground, just as frigate birds snatch hatchling turtles from sandy beaches, or scooped fishes from the surface waters of rivers, lakes and oceans, as do skimmers. (These birds fly with their lower mandibles immersed, and seize any prey contacted as they do so.)

A typical pterosaur, such as *Rhamphorhynchus phyllurus* (Fig. 33), had a characteristic archosaurian skull born on a long, flexible neck, with a continuous series of ribs between the pectoral and pelvic girdles. Two separate groups of pterosaurs are recognized: the Pterodactyloidea (Fig. 34), in which the tail is short, and the Rhamphorhynchoidea, which had long, straight tails up to twice the length of the backbone in front of the pelvis (Fig. 33). Impressions in the rock indicate that they had a flattened membrane at the end, which might have been of assistance in steering. The first three fingers were reduced to small hooks, which probably served as hangers for roosting in trees or on cliffs, while the fifth finger was lost. Projecting forward from the wrist was an elongated pteroid bone which helped to support the wing membrane. Although it did not project forwards like that of birds, the sternum was enlarged and served for the attachment of the large pectoral muscles that moved the wings.

During the Jurassic period there were several species of rhamphorhynchoids, but, in Late Jurassic times these were supplanted by the pterodactyloids in which the tail was suppressed and the dentition reduced. In some of the most advanced members of that group, teeth were completely absent and the jaws took the form of bird-like beaks. This may have reflected a change in diet from flesh to smaller items, but probably occurred partly in response to the need for a reduction in weight, especially at the head which is rather far from the centre of gravity. The function of the teeth in birds has been transferred to the gizzard, the position of which is such that its weight does not unbalance the body.

Escape from predators and the capture of prey, as already mentioned, are undoubtedly the two aspects of natural selection that have most influenced the morphology and lifestyles of amphibians and reptiles. Which of them has been the more important in the evolution of the ability to jump, glide or fly cannot easily be determined because these three endowments have multiple functions and are used both in defence and offence. Certainly, however, nobody who has studied the behaviour of modern amphibians and reptiles in the field would disagree with the opinion that their most conspicuous environmental adaptations have been engendered in response to predation.

## 5.4 Locomotion in Water

In their postcranial skeleton, as we have seen, the Ichthyostegalia showed a mixture of piscine and amphibian characters. Their vertebrae retained essentially crossopterygian features, and the fin rays of the fish tail were still present, although there were strong pectoral and pelvic girdles and well-developed limbs. Presumably they swam like fishes and walked like urodeles. In the Sirenidae, the pelvic girdle and hind limbs are absent, while the fore limbs are very much reduced in size. Living sirenids are restricted to the south-eastern United States and north-eastern Mexico, but fossil

Sirenidae were widespread in North America with one genus *Habrosaurus* in the Upper Cretaceous (Duellman and Trueb 1986). In these aquatic, eel-like salamanders, which constitute a distinct suborder, the loss of legs is not related to burrowing behaviour as it is in the caecilians, but to an aquatic mode of life.

The swimming movements of urodeles are essentially the same as those of fishes. Waves of muscular contraction pass down the body, engendered by the axial musculature of the vertebral column. Figure 25 illustrates how the inner sides of the sinuous curves of a snake press against projections when it is moving over a smooth board. When a fish is swimming, the inner sides of the curves of its body press, in a similar way, against the water. Resistance is increased by expansion of the dorsal, anal and caudal fins. In most eels, the anterior part of the body is cylindrical in cross section, whereas the posterior part is compressed. This provides mechanical advantage like the expanded caudal fins of other fishes. The flattened portion of the body is given additional purchase by a continuous, unpaired fin which combines the dorsal, anal and caudal fins of most other fishes. This unpaired fin extends from the first third of the back, round the end of the tail and forward as far as the anus, which is situated a little in front of the middle of the body (Fig. 35).

The tail of a larval newt or salamander, like that of an eel, is an adaptation for swimming. The males of all the British newts (*Triturus* spp.; Salamandridae) have a cutaneous crest on the back and tail during the breeding season. This develops soon after they return to the water in the spring, and is absorbed almost completely after the breeding season has ended. In those individuals which remain in water throughout the year, the development of the crest begins in autumn. Crests differ in size and shape in different species. They are rich in sense organs and, although they lack muscles, they do assist to some extent in swimming. Nevertheless, their main function is for display during courtship (Chap. 8; Smith 1951). Female newts and salamanders which lack crests naturally do not swim quite so efficiently because their bodies do not obtain good purchase against the water, but the functional significance of

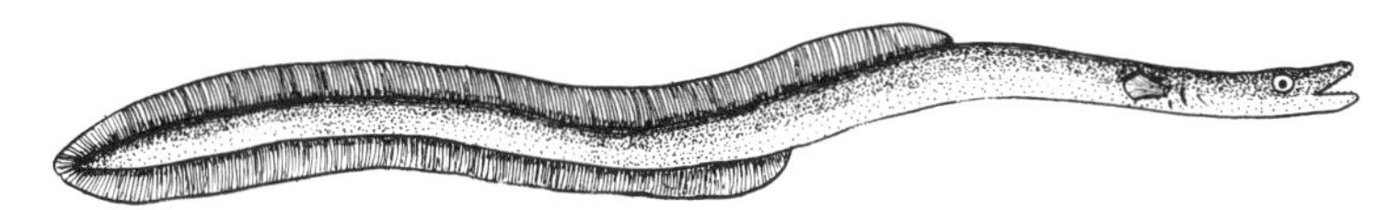

**Fig. 35.** Lateral view of a swimming eel to show the fins. Each half wave length of the body represents one propulsive unit. See text for further explanation

the dorsal crest of male European crests remains observe, despite intensive research into their sexual behaviour (Halliday 1992).

There is nothing subtle about the swimming of frogs and toads apart from the fact that, in the African clawed toad (*Xenopus laevis*), the Surinam toad (*Pipa pipa*) and some species of *Hyla*, the feet are webbed. This gives additional purchase when the limbs are extended, as by a human being swimming the breast stroke. The male European toads *Bufo bufo* and *B. viridis* exhibit a slight extension of the webbing between the toes during the breeding season when the animals return to water. This involves growth of the toe and tarsal ridges to form fringes in the male *Pelodytes punctatus*, and in *Elosia* and *Crossodactylus* spp. Toe webs occur only sporadically in one or two species of very distinct families of Anura (Noble 1931).

Geese and swans paddle with their feet on the surface of the water when they take flight, but not many vertebrates are able to walk or run on water without the assistance of wings. This seemingly impossible feat is, however, achieved by certain lizards belonging to the genera *Anolis, Crotaphytus, Basiliscus, Lacerta, Eumeces* and *Hydrosaurus*, whose feet are often adapted with enlarged, flattened scales as in the case of *H. pustulosus* (Fig. 18C). When *Basiliscus basiliscus* (Fig. 22) is running bipedally across water its feet slap down, parallel to the surface, with such speed and pressure that the surface is forced downward, creating an upward force of equal strength which supports the weight of the body. After impact, the leg is pushed further downwards and backwards so rapidly that a cavity is produced around it and the water surface is not broken. Before this cavity collapses, the lizard's foot is pulled upwards ready for the next step. A basilisk moves at about 8 km/h. An animal weighing as much as a human being would need to run at 108 km/h or more across the surface of a stretch of water to avoid sinking. Yet an Olympic sprinter only reaches a maximum of about 43.5 km/h!

During the Mesozoic era, several groups of reptiles became adapted to marine life. Foremost among them were the Ichthyosauridae (Fig. 36) which first appeared in the Middle Triassic. Fish-like in shape, they were highly modified for aquatic life, and swam as fishes do. Their tails were specialized for propulsion, legs flattened into paddles for balancing, and with a fleshy dorsal fin which prevented their bodies from rolling (McGowan 1991). Their jaws were elongated with numerous teeth for catching the fish on which they fed. Although ecological analogues of porpoises and dolphins, they moved their tails from side to side as fishes do, not up and down like the flukes of a whale. Their ancestry is little known (Colbert 1965). McGowan

**Fig. 36.** Reconstruction of an ichthyosaur (*Ichthyosaurus* sp.) (length ca. 2.5 m), Upper Jurassic. (Cloudsley-Thompson 1994)

(1991) discusses their mechanics of swimming with special reference to ichthyosaurs.

If the ichthyosaurs were analogous to porpoises and dolphins, the nothosaurs, which reached their peak in Late Triassic times, were the equivalents of seals and sea lions. The suborder Nothosauria comprised elongated reptiles with large paddles used for swimming and long, sinuous necks. Their skulls were small and equipped with sharp, piscivorous teeth. They probably lived at the edges of the oceans, pursuing fishes in shallow water, perhaps even scrambling onto land to bask in the warm sunshine of the Late Triassic period when they reached their peak. Their descendants, the plesiosaurs, were very much larger - *Elasmosaurus* sp. (Fig. 37) was over 12 m long - and inhabited the oceans of the world from the Middle Triassic until the end of the Cretaceous period, long outliving the ichthyosaurs. Some of the plesiosaurs were long-necked elasmosaurs with relatively small skulls; others were short-necked pliosaurs (Fig. 38) with long skulls and jaws. The former probably swam on the surface catching squids and fishes by darting their snake-like necks to one side or the other, while the pliosaurs chased fishes through the water, as did the ichthyosaurs. It is not impossible that the ichthyosaurs may have died out as a result of competition with pliosaurs.

The short-necked pliosaur *Stretosaurus macromeros* of the Upper Cretaceous was probably the largest marine reptile of all time. Its mandible was over 3 m long, so the total length of the body must have been more than 15 m. *Kronosaurus queenslandicus* (Fig. 38) was of comparable size. A complete

**Fig. 37.** Reconstruction of a long-necked plesiosaur (*Elasmosaurus* sp.) (length ca. 12 m), Upper Cretaceous. (Cloudsley-Thompson 1994)

**Fig. 38.** Reconstruction of a pliosaur (*Kronosaurus* sp.) (length ca. 15 m), Upper Cretaceous

skeleton in the Museum of Comparative Zoology, Harvard University, measures 12.8 m in length. It is now known that the pliosaurs had a fin on the upper part of the tail and were much more streamlined than previously realized. They would have been formidable competitors of the ichthyosaurs, whose average size was about 2 m, although *Leptopterygius acutirostris* of the Lower Jurassic may have reached 13 m some hundred million years earlier.

In addition to squids, belemnites (a group of fossil molluscs thought to be ancestral to the modern cephalopods excluding nautiloids) and fishes, the food of the elasmosaurs may possibly have included flying animals such as pterosaurs and toothed sea birds. The stomach contents of some fossils have been preserved and include the remains of these creatures. In addition, up to several hundred stomach stones and gastroliths have been found in the belly region of a single plesiosaur. At one time it was thought that the stomach stones of plesiosaurs helped in grinding the food. The presence of fragile but unpulverized bones and shells along with the stones belies this, however, and it is now believed that the stones that are swallowed serve to neutralize buoyancy. Crocodiles also swallow pebbles as ballast so that they can lie hidden below the surface of the water with only the eyes and nostrils protruding above the surface. The weight of these stomach stones also helps them to drag large prey animals underwater so that they are drowned (Cott 1961).

In Early Triassic times, the Placodontia (plate teeth), contemporaneously with the Nothosauria, also became specialized for marine life, but in shallow water. These massive reptiles had stout bodies, short necks and tails and paddle-like limbs. In some genera of placodonts, such as *Placochelys* and *Henodus* (Fig. 39), the body was very broad and armoured with bony plates, giving an appearance strikingly similar to that of the marine turtles of today, although they were not related to them: turtles are Anapsida, placodonts were Synaptosauria. Their lifestyle, too, was probably more like that of walruses than of turtles. Their large, flattened plate-like teeth were obviously adapted for crushing shellfish and they probably dived to the bottom of shallow waters to dig up clams and other molluscs. In some genera, including *Henodus*, the teeth have disappeared altogether and, as in chelonians, the jaws were probably covered with horny plates. Placodonts and nothosaurs were almost certainly secondarily aquatic and shared a protorosaurian ancestry. [The order Protorosauria (dawn lizards) comprised an odd collection of possibly unrelated Permian and Triassic reptiles, some lizard-like in appearance, others with small limbs, elongated bodies and extremely long necks; (Bellairs and Attridge 1975).]

**Fig. 39.** Reconstruction of a placodont (*Henodus* sp.) (length ca. 1 m), Lower Triassic. (After Špinar and Burian 1972)

Among the earliest of known reptiles and the first aquatic types were the mesosaurs (e.g. *Mesosaurus*; Fig. 40) of the Pennsylvanian period. Despite their early appearance, they were highly specialized, with long, slender jaws armed with sharp piscivorous teeth and broad tails clearly adapted for swimming. Possibly inhabitants of freshwater, their relationships are not clear. A.S. Romer (Romer and Parsons 1986) advocated placing them within the subclass Anapsida, but other authorities now regard them as synapsids.

Although the mid-continental seas of Late Cretaceous times saw the disappearance of the ichthyosaurs, they were enriched not only by elasmosaurs and pliosaurs, but also by Mosasauria – gigantic predatory marine lizards, distantly related to the Varanidae of today. One of these, *Tylosaurus* (Fig. 41), reached a length of 9 m. The mosasaurs were by no means the only aquatic Mesozoic lizards: others included the Dolichosauria and Aigialosauria, small, long-tailed amphibious lizads with limbs in varying degrees converted into paddles. The aigialosaurs were the more primitive and probably ancestral to the mesosaurs. All three families are included in the infra-order Platynota. Finally, mention should be made of the geosaurs or thalattosuchians, members of a family of sea-going Jurassic crocodilians. In these, a heterocercal tail evolved, with a sharp downward bend of the caudal vertebrae, while the limbs were modified to form paddles (Colbert 1980). Although usually regarded as Eusuchia, these too may have been aberrant lizards.

**Fig. 40.** Reconstruction of a mesosaur (*Mesosaurus* sp.) (length ca. 75 cm), Pennsylvanian – upper subperiod of the Carboniferous

**Fig. 41.** Reconstruction of a mosasaur (*Tylosaurus* sp.) (length ca. 9 m), Upper Cretaceous

The bones of the wrist, which are cartilaginous in salamanders, were firmly ossified and closely united in plesiosaurs. This can be understood in relation to the fact that salamanders crawl on underwater surfaces or swim with their tails whereas plesiosaurs, like turtles, swam with their fins. Another adaptation to life in water is seen in the eye. The accommodation of the eye in reptiles takes place through the compression and elongation of the eyeball by means of external muscles. The imbricated sclerotic plates permit expansion and contraction of the eyeball. The cornea changes its contour unless some compensatory force counterbalances it: in aquatic reptiles the sclerotic plates provide this. Among terrestrial reptiles such as skinks, and the dinosaurs *Diplodocus* and *Trachodon*, the sclerotic plates are ossified, and strong sclerotic plates appear to be a necessary adaptation for aquatic life (Williston 1914).

Modern reptiles show varying degrees of adaptation to an aquatic lifestyle. In amphibians, freshwater terrapins (Emydidae) and snappers (Chelydridae), the toes are moderately or completely webbed, although they usually possess well-developed claws. The soft-shelled freshwater turtles (Trionychidae) are more advanced in that their feet, especially the front ones, are roughly paddle-shaped, with huge webs and claws restricted to the three inner digits: they can swim quite fast too. Freshwater terrapins and turtles row themselves along by lateral strokes of the limbs which operate in diagonal pairs, as in walking. The hind limbs is usually longer and more powerful than the fore limbs, which in the genera *Graptemys* and *Pseudemys* are held against the sides of the body when swimming, while the hind legs kick alternately.

Sea turtles, such as the leatherback (*Dermochelys coriacea*), hawksbill (*Eretmochelys imbricata*) (Fig. 42) and green turtle (*Chelonia mydas*), are even more specialized for aquatic locomotion. Their bodies are streamlined and the limbs, although still pentadactyl (five-fingered), are transformed into true paddles. The digits are connected throughout their length, and the claws reduced to one or two on each limb – in the adult leatherback, even these have disappeared. The front paddles are longer than the hind ones and are the main propulsion organs. They drive the body with almost vertical strokes, like those of a bird's wing, so that the animals seem to fly through the water. They are used together, not alternately, and the back limbs are employed mainly as rudders (Bellairs 1969). The nothosaurs and plesiosaurs probably swam in a similar way.

Although fully adapted for oceanic life, even sea turtles nest on land. This double habitat imposes the need for them to travel, and some species may

**Fig. 42.** Hawksbill turtle (*Eretmochelys imbricata*) (length ca. 1 m). (Cloudsley-Thompson 1994)

swim for thousands of miles before migrating back to the very same beach on which they, themselves, hatched many years before. For instance, the herbivorous green turtle commutes every 2 or 3 years between the expansive stands of 'turtle grasses' (*Thalassia* and *Zostera* spp.) which carpet the protected, shallow waters of their feeding territories and distant nesting beaches where huge waves thunder unhindered onto the sandy shore. The conversion of a terrestrial limb into a paddle involves lengthening its distal segment. In the ichthyosaurs, plesiosaurs and mesosaurs, this was achieved by increasing the number of phalanges in each digit. Moreover, in the ichthyosaurs the number of digits was also increased. In marine chelonians, however, the paddles have evolved by elongation of the individual bones, especially the metacarpals and phalanges of the hand, and there has been no multiplication of their number. The typical phalangial formula of the hands of most freshwater species (2:3:3:3:2 or 3) has been preserved. Four phalanges are found in the Trionychidae, but this probably represents a primitive reptilian condition rather than an addition to the original chelonian number (Bellairs 1969).

The variety of gaits of crocodilians on land has already been described: in water they use their feet as paddles when moving slowly, but propel themselves mainly with their flattened tails when swimming fast. The principles involved are essentially fish-like, as were those of the ichthyosaurs.

Large crocodiles can withstand seawater for considerable periods of time because their bodies have a relatively large surface to volume ratio. They can swim remarkably well with their flattened fish-like tails. The estuarine crocodile (*Crocodylus porosus*) is sometimes seen far out to sea, but, like the American *Crocodylus acutus*, it is mainly an estuarine species and breeds on land like all other crocodilians as well as marine turtles and freshwater terrapins.

The only lizard which has a claim to be genuinely aquatic is the Galapagos marine iguana (*Amblyrhynchus cristatus*), which lives among rocks on the sea shore and enters the sea to graze on algae. Although it dives quite deeply when feeding, it cools rapidly in water and quickly returns to the warmth of the land.

Some sea snakes (Hydrophiidae) are the only modern reptiles to live and breed completely at sea, without ever returning to land. These all belong to the subfamily Hydrophiinae. In conformity with their completely aquatic life, they are ovoviviparous. In contrast, members of the subfamily Laticaudinae are amphibious, and are able to move about on land. Although they retain many features more typical of terrestrial forms, including well-developed ventral scales, like the Hydrophiinae they also have flattened paddle-shaped tails (Fig. 43) and nasal valves, and swim as eels do. (Incidentally, the presence of a single row of wide ventral scales or gastrosteges, which are important in locomotion on land, distinguishes most snakes from legless lizards in which the ventral scales are always small like those on the back and sides.)

It is surprising that sea snakes have few special physiological adaptations for diving, although some species are reportedly able to remain submerged

**Fig. 43.** Sea snake (*Emydocephalus annulatus*) showing flattened tail (length ca. 7.5 cm). (From a photograph)

perhaps for up to 8 h, and dive to depths of 100 m or even more. Their oxygen consumption is similar to that of land snakes. They share with these, and with freshwater turtles (Girgis 1961), the capacity for cutaneous gaseous exchange. They also voluntarily undergo long periods of apnea (holding their breath). Alternation of breathing periods with apnea also occurs in some terrestrial lizards, however, and the survival time of sea snakes breathing pure nitrogen is much less than that of most freshwater turtles. Furthermore, their survival in oxygen-free water is as low as 21 min, which is even shorter than that of most terrestrial lizards and snakes, or of crocodilians. The change in the heart rate of sea snakes during diving is similar to that of other diving reptiles and of terrestrial snakes forcibly submerged, while the haematological characters of their blood are similar to those of land snakes. Indeed, Heatwole and Seymour (1975) concluded that sea snakes do not have any unique physiological characteristics related to their diving habits, but have merely extended some of them to a degree not usually found in their terrestrial relatives (see also Seymour 1982).

# 6 Diversity of Anti-predator Devices

Predation is a constant threat to all except the largest of animals, and even among these only adult representatives of forms such as giant tortoises, crocodiles and elephants are virtually immune to attacks from enemies other than Man. Most living amphibians and reptiles occupy intermediate positions in the food chains of which they form a part. The majority are themselves carnivorous and therefore in the ambiguous situation of being simultaneously both predators and prey. It is with their adaptive responses to the latter role that the present chapter is concerned.

The early stages in the lives of amphibians and reptiles are especially vulnerable to predatory invertebrates, but, when they grow larger, vertebrates are the main enemies of both. Tadpoles and larval salamander are regularly preyed on by dragonfly larvae, Heteroptera, water beetles and other predaceous aquatic insects, and amphibian eggs by arthropods and leeches – although bacteria and fungi may cause even higher mortality (Zug 1993). In permanent waters, not unexpectedly, fishes are the major predators of amphibian larvae. On land, small amphibians and reptiles are often captured by centipedes, scorpions, spiders and other carnivorous arthropods: as they get bigger, their principal enemies are larger arthropods and reptiles, as well as birds and mammals. Wading birds in particular prey heavily upon larval and adult amphibians when these are in water. They also eat water snakes, baby crocodiles (Fig. 44) and small terrapins and turtles. On land, raptors catch large numbers of lizards and snakes, while secretary birds (*Sagittarius serpentarius*) and road runners (*Geococcyx* spp.) specialize on such reptiles. Mammals are far less significant as predators, although mongooses and felids play part in controlling reptile populations.

Anti-predator devices that have evolved in response to one type of predator are not always effective against another. Concealment may be not only a defensive response to predation by other animals but also an aggressive

**Fig. 44.** Young Nile crocodile being devoured by a saddle-bill stork. (Cloudsley-Thompson 1994)

adaptation, enabling its possessor to approach its own prey without being observed.

The defensive strategies of animals are of two kinds, primary and secondary. The former are defined as those which operate regardless of whether a predator is in the vicinity or not, whilst secondary defences are invoked only when the prey has been detected by a potential predator.

## 6.1 Concealment

Although most amphibians appear to be rather defenceless creatures and, indeed, are devoured by a great variety of predators, they are usually by no means lacking in protective devices. One of the most important of these is cryptic or concealing coloration. Many amphibians and reptiles have colours and patterns that tend to match those of the substrates on which they exist. The coloration of many amphibians (and of some reptiles) can be extremely variable. This variation is so extreme at times that an almost spotless form of the American leopard frog (*Rana pipiens*) was once thought to be a different species. It is known to differ from the common form only by a single gene. The same thing occurred in the case of the Venezuelan tree frog *Gastrotheca*

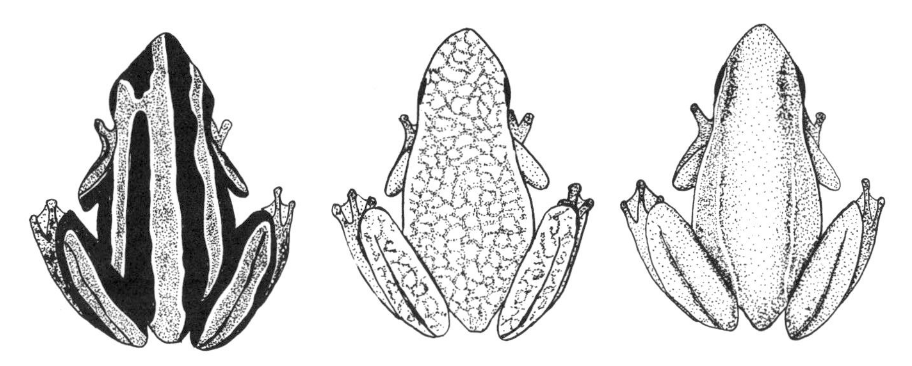

**Fig. 45.** Polymorphism: striped, speckled and plain individuals of the tree frog *Hyperolius marmoratus*. (After Mattison 1987)

*monticola*. In a number of species, the males and females are coloured quite differently (Frazer 1973). Polymorphism, in which a species occurs in a number of different forms, is illustrated by a number of species of small frogs, such as the African reed tree frog (*Hyperolius marmoratus*) (Fig. 45; Mattison 1987).

In many amphibians, and some reptiles too, crypsis is enhanced by colour change. The colour of the skin depends upon a number of pigment cells or chromatophores which are scattered through it at different depths. The most superficial of these are sometimes known as lipophores because they contain droplets of orange or yellow fat. They are more usually called xanthophores or erythrophores, however, and are responsible for the yellow, orange and red colours of the skin. The pigments that impart these colours are carotenoids and pteridines. Below them are light-reflecting iridophores. These are sometimes called guanophores because they contain blue or white crystals of the purine guanine (the excretory compound formed by the breakdown of uric acid). The deepest pigment cells are melanophores, which contain black, brown or red melanin in subcellular orangelles or melanosomes (Fig. 46). The apparent colour of any particular piece of skin depends on not only the relative numbers of the different types of chromatophore in it, but also the degree to which these pigment cells are expanded or concentrated. A further complication lies in the fact that, although the colours seen are partly the result of the chemical substances in the chromatophores, the reflection and refraction of light are also involved. Green colours appear when blue rays, reflected by iridophores, are filtered through the yellow pigment of the

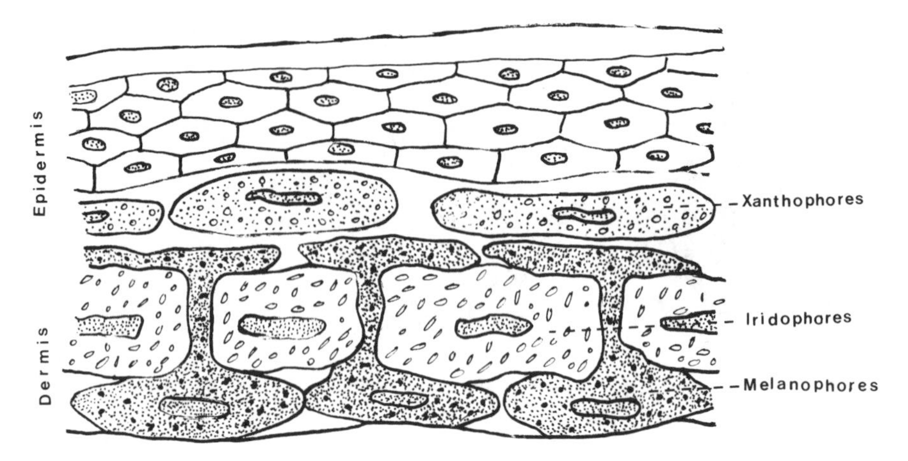


**Fig. 46.** Transverse section through the skin of an amphibian showing the chromatophores (much enlarged)

overlying xanthophores (Frazer 1973; Herman 1992; Stebbins and Cohen 1995).

Chromatophores are found throughout the body, including the blood vessels, peritoneum, mesentery and skin. They not only provide protective as well as epigamic or sexual coloration, but also assist in thermoregulation. Dermal melanophores are star-shaped and participate in rapid physiological colour change, either through dispersion (which causes darkening) or their aggregation (which causes lightening) of the melanosomes. Two kinds of colour response are found in amphibians: primary blanching of the larval stages in darkness, and the environmental adaptation by which the animals match their surroundings. In both cases, the hormones that control the reaction are released into the bloodstream and act directly on the melanophores. The physiology of colour change in amphibians has been reviewed by Bagnara (1976). (For an account of recent research on the amphibian integument, see the articles in Heatwole and Barthalmus, 1994.)

Like those of amphibians, the colours of reptiles depend to a large extent upon the presence of chromatophores, mostly situated in the outer regions of the dermis, and these operate in the same way in both groups. The melanophores usually form a dense, almost continuous subepidermal layer; xanthophores lie outside them, immediately beneath the epidermis, while between the two are to be found iridophores containing semi-crystalline particles of guanine. Although the hues of most reptiles are constant, their patterns are often built up from a mosaic of differently coloured scales. Several species are able to change colour so that they retain their crypsis in different environments. Examples include chameleons, *Anolis carolinensis* (Iguanidae) and the Asian agamid *Calotes versicolor*.

As in amphibians, reptilian pigment cells are not confined to the skin and, in lizards, the peritoneum is sometimes deeply pigmented due to an abundance of melanophores. The suggestion has been made that this may protect the alimentary canal from harmful ultra-violet radiation. In certain species of *Anolis*, a correlation has indeed been found between such pigmentation and the habitat in which the lizard normally occurs. Species in which pigmentation is slight live in shady woodland, while those in which the pigmentation is more extensive inhabit open country and are exposed to greater intensities of radiation (Colette 1961). In other reptiles no such correlation has been found and, moreover, the skin of some species is impervious to UV radiation of the shorter wavelengths (Bellairs 1969).

Concealment in amphibians and reptiles is not achieved by general colour resemblance alone, as Cott (1940) has emphasized. Many species show countershading and are more pale on their ventral than on their dorsal surfaces, thereby eliminating the effects of shadow. Elimination of shadow is also effected by close application of the body to the substratum. For example, a green tree frog, pressing itself against a leaf of the same colour, may be almost invisible to the human eye. The same applies to geckos tightly clinging to bark or rocks whose coloration is closely matched by their own.

There are three fundamental principles of concealment: colour resemblance, countershading and disruptive contrast. The last of these is especially well developed among amphibians and reptiles. The maximum effect is produced when tones of high contrast are adjacent to one another so that they draw attention away from the outlines of the body of the animal concerned. The principal is superbly illustrated in Cott's (1940) *Adaptive Coloration in Animals*. Many frogs, belonging to widely separated families, have the sides of the head coloured with dark, irregular blotches which serve to camouflage the eye and break up the outline of the body. This phenomenon is seen in the common European *Rana temporaria*, many other species of *Rana*, *Cardioglossa leucomystax*, *C. gracilis*, *Hyla savignyi*, *H. regilla*, *Polypedates leucomystax* and so on, whose disruptive coloration hinders recognition of their possessors for what they are. Similar disruptive effects are found in many lizards such as *Mabuya* spp. which have dark lateral stripes that run through the eyes. On account of their circular shape and black pupils, the eyes of vertebrates are inherently among the most conspicuous and easily recognized of natural objects. It is therefore to be expected that they should be well camouflaged in cryptic animals, which include numerous amphibians, lizards and snakes (Cloudsley-Thompson 1994).

Not only the eyes but also the limbs can present a familiar appearance to predatory enemies. This may be overcome by coincident disruptive colour patterns. For instance, *Tachycnemis* (= *Megalixalus*) *fornasinii* (Hyperoliidae) from the Seychelles islands has dark markings on its body and limbs which, when the frog assumes its natural attitude of rest, break up its form into strongly contrasted areas of light and dark colour (Fig. 47). The same principle is invoked, as the result of parallel evolution, in the disguise of *Afrixalus quadrivittatus*, a Kenyan hyperoliid and of the South American tree frog *Hyla leucophyllata* (Hylidae). In the tree frog *Rhacophorus fasciatus* (Polypedatidae) from Sarawak, the toad *Hylodes longirostris* (Leptodactylidae) from Ecuador, and various other anurans, the coincident disruptive patterns are horizontal and run across the body. The same is true of numerous reptiles, especially snakes (Cloudsley-Thompson 1994).

Coincident disruptive coloration is usually not so clearly displayed by reptiles as it is by amphibians. Concealment of the eye, however, occurs frequently in both groups, as already mentioned. In many tree frog, nocturnal geckos, and snakes such as desert vipers, as well as in crocodilians, the pupil becomes a thin slit when contracted in daylight. Not only is this an effective means of protecting the ultra-sensitive retina – a slit pupil can exclude even more light when closed than can a circular pupil – but the colour of the iris is

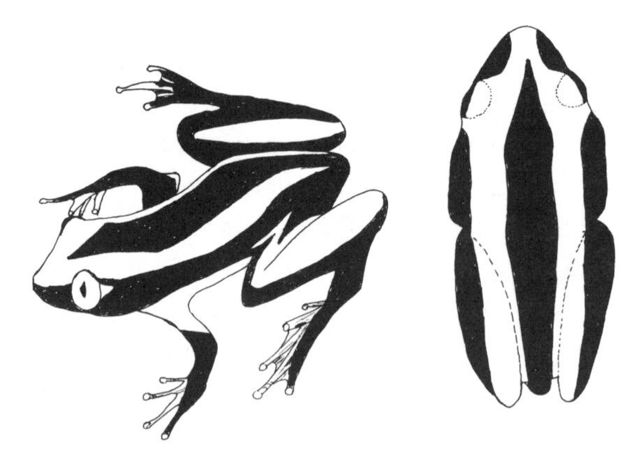

**Fig. 47.** Tree frog (*Tachycnemis fornasinii*), drawn to illustrate the principle of coincident disruptive coloration. (After Cott 1940)

the same as that of the remainder of the head and body, so that the eye is no longer conspicuous (Cott 1940).

Coincident disruptive eye marks are found in many amphibians and reptiles – frogs and snakes in particular – which not only conceal their possessors from enemies, but also help them to surprise their prey. Black horizontal stripes from the posterior of the eye to the tip of the snout in American vine snakes (*Oxybelis* spp. and *Uromacer* spp.) (Colubridae) aid vision by assisting the snakes to line up their heads with the prey. The eye line may be extended by protruding the tongue; in some species it may be extended for as long as 20 min. In the African twig or bird snake (*Thelotornis kirtlandii*) (Colubridae) (Fig. 29), the markings are even more of an eye mask. Whereas *Oxybelis* and *Uromacer* spp. have round pupils, which are typical of most day-active colubrids, *Thelotornis* and long-nosed whip snakes *Ahaetulla* (= *Dryophis*) spp. of south-eastern Asia have horizontal, key-hole pupils. Other examples of eye stripes are to be found in Boidae such as *Boa* (= *Constrictor*) *constrictor* and *Python molurus*, in Viperidae including *Agkistrodon* spp. and *Vipera* spp., as well as in *Gastropyxis smeragina*. Yet another effective way of concealing the eye involves the use of large patches of dark tone spreading over the entire orbit. Crypsis among Amphibia is mentioned in passing by a number of authors, but has not been reviewed in such detail as it has in reptiles by Greene (1988).

## 6.2 Advertisement

With few exceptions, only very large, formidable or distasteful animals can afford to be conspicuous in a world in which smaller forms are usually beset by predatory enemies. If they are toxic or venomous, however, the evolution of conspicuous, coloration may well prevent potential enemies from attacking or harming them inadvertently, before they have been recognized. Warning or 'aposematic' colours are of value in drawing attention to the fact that an animal possesses an effective means of defence and is best left severely alone (Cott 1940; Edmunds 1974; Cloudsley-Thompson 1980, 1994).

Aposematic coloration is invariably achieved by the use of only a few striking colours and bold patterns. The fewer the colours and configurations that are employed, the easier it becomes for unpalatable prey to be recognized. Moreover, a common type of recognition signal, as is displayed by a wide variety of aposematic forms, greatly facilitates learning. That is why so many aposematic animals have similar combinations of revealing hues - red, black and white; orange and black; yellow and black; yellow, black and white. These are the colours that are most striking when seen against the normal green and brown surroundings of nature. It must be remembered, however, that not all conspicuous colours are aposematic. Some vividly coloured amphibians and reptiles may be cryptic in their natural environments. Brilliant-green tree frogs may be quite invisible when crouching on a leaf in tropical forest. The bold design of the gaboon viper (*Bitis gabonica*) (Fig. 48), so conspicuous in museums and reptiliaries, in fact comprises disruptive patterns which render the snake inconspicuous in the normal rainforest environment of dark shade with sun flecks, green and yellow leaves, black bark and grey lichen.

The yellow-bellied sea snake (*Pelamis platurus*) (Hydrophiidae) is very conspicuous when not floating in the debris of drift lines, its distinctive tail being especially noticeable. This aposematic coloration is associated both with venom and with distastefulness. In the eastern Pacific, *P. platurus* has no known aquatic enemies, and predatory fish, such as snappers, refuse even to nibble at it unless it is experimentally camouflaged in a piece of squid; and then they reject the morsel as soon as they have tasted it. When offered to predatory Atlantic fish that have never before encountered a sea snake, however, it may be eaten and, in about one of 12 meals, the predator dies (Rubinoff and Kropach 1970). It may, however, benefit sea snakes not to be visible all the time. Although *P. platurus* seems to be remarkably free from predation, sea snakes are nevertheless preyed upon by sea eagles and other marine birds (Heatwole 1975) and have also been taken from the stomachs of Philippine moray eels and sharks.

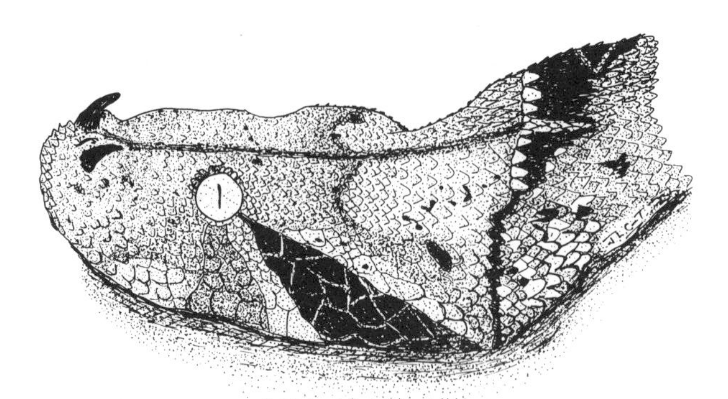

**Fig. 48.** Head of gaboon viper (*Bitis gabonica*). The disruptive black triangular patch disguises the mouth. (Cloudsley-Thompson 1994)

Aposematic colours are usually combined with sluggish behaviour and free exposure, as in the case of Helodermatidae (see below). Although the majority of Anura have a cryptic, patterned coloration in which greens, browns and greys predominate, a few species are strikingly coloured with conspicuous, saturated hues which are almost as brilliant as those of tropical insects and birds. However, unlike many of the latter, frogs and toads are poorly equipped for escape from potential predators – especially snakes and birds which depend largely upon vision when hunting prey. Yet the dart-poison frogs (Dendrobatidae) are among the most conspicuous animals in the world and, at the same time, the most toxic. In Colombia, Indian hunters poison their blowgun darts with extremely poisonous alkaloids secreted by these small amphibians (Myers and Daly 1983). Unlike most other frogs, dendrobatids are active and conspicuous during the day, foraging amongst the leaf litter of tropical rainforest, and moving with short leaps. They are rarely still for more than a second or two so that they cannot fail to attract attention. Although cryptic coloration may, or may not, be associated with noxious properties, aposematic coloration is characteristically associated with an effective means of defence. Any conspicuous amphibian or reptile that does not possess toxins or venoms is almost invariably the mimic of a species that does.

Among the Urodela, the European fire salamander (*Salamandra maculosa*) possesses a virulent toxin which is advertised by warning colours and sluggish behaviour, while the only venomous lizards, the North American Gila monster (*Heloderma suspectum*) (Fig. 49) and the Mexican beaded lizard (*H. horridum*) (Helodermatidae), provide striking contrast to most other desert lizards which depend for safety upon cryptic coloration, alertness and speed. Not only are the venomous species brightly coloured, but also they are slow, defiant, and defend themselves with their powerful jaws and poisonous

**Fig. 49.** Gila monster (*Heloderma suspectum*) (length ca. 60 cm). (Cloudsley-Thompson 1994)

bite. The bright colours of some aposematic amphibians are normally concealed. Only when the animal is threatened are the colours suddenly revealed. For instance, the froth-nest frog *Pleurodema brachyops* (Leptodachylidae) of Central and South America displays by inflating its body and elevating its rear to reveal paired 'eye spots', one in each upper groin and on the back of the thighs (Martins 1989). A similar eye spot display is exhibited by the Brazilian toad *Physalaemus nattereri* (Leptodactylidae) (Fig. 50) when it presents its rump towards predators (Edmunds 1974). (Startling or 'deimatic' behaviour will be discussed below.)

The majority of venomous snakes are cryptic. When bright colours occur, more often than not they form coincident disruptive patterns and enhance concealment. Conspicuous forms, such as coral snakes (and their mimics), are somewhat exceptional. Although crypsis in venomous snakes may have an offensive function, in that the animals are not readily seen by their prey, aposematic coral snakes (*Micrurus* and *Micruroides* spp.) (Elapidae) apparently do not experience difficulty in capturing their food. As I have argued in greater detail elsewhere (Cloudsley-Thompson 1994), it does not benefit snakes to be profligate with their venom, which is used more profitably in the subjugation of prey than in defence against potential predators. For some reason, venomous animals in general tend to be cryptic while many toxic forms are aposematic: but, once a venomous snake has been detected, it will still be advantageous to it if, as we shall see, the enemy can be frightened off before making an attack that may harm the snake even if the latter manages to kill its adversary.

Deflection marks which direct attacks away from vulnerable parts of the body can have the disadvantage that they are conspicuous and may actually

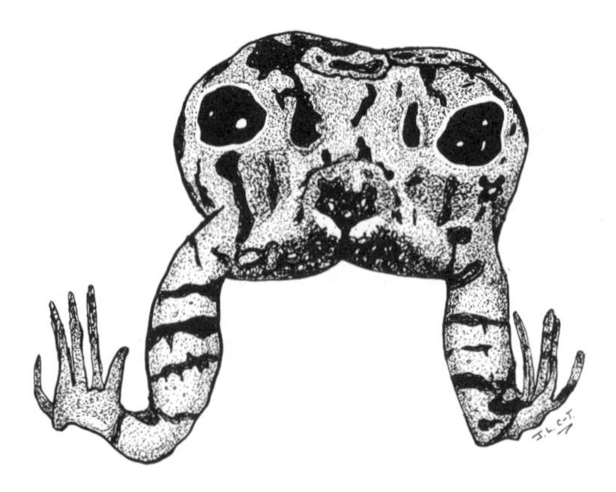

**Fig. 50.** Defensive eye spot display of the toad *Physalaemus nattereri*

attract the attention of a predator which had not previously noticed the animal (Edmunds 1974). On balance, however, they must be beneficial or they would not otherwise have evolved. The tails of certain lizards are, likewise, sometimes more colourful than the rest of the body, and are brighter in the young than in the adult. For instance, the tail of *Mabuya quinquetaeniata* (Scincidae) is bright blue in contrast to the black body stripes, but this tail coloration disappears with advancing age. In South Africa, *Eremias namaquensis* (Lacertidae) has a red tail when young: this, too, becomes cryptic later in life (Cott 1940). Lizards often sacrifice their tails in autotomy and it is better for them that the tail should be attacked by a predator than that their heads should suffer. In *Tiliqua rugosa* (Scincidae) the stumpy tail is very similar in shape and colouring to the head. Automimicry of the head by the tail is also found in some geckos, uropeltid snakes, and in amphisbaenians. Not only is the attention of prey distracted from the menace of the true head, but also the attacks of predators are diverted to a less vulnerable part of the body.

There is no means of knowing the colours in life of extinct amphibians and reptiles, but the same principles were probably involved as with modern forms. Smaller species would have been cryptic unless the adults exhibited bright sexual or 'epigamic' coloration during the mating season. Larger carnivorous dinosaurs may well have been offensively cryptic, being mottled, striped or spotted, like the larger Felidae which fill a comparable ecological niche today. The larger herbivores were probably more uniformly coloured, as are elephants and rhinoceroses (Cloudsley-Thompson 1994).

## 6.3 Mimicry

Distasteful, toxic and venomous animals are normally avoided by potential predators, so it is not surprising that many harmless species should, as a result of natural selection, have come to resemble them. This phenomenon, named 'Batesian mimicry', was first described in 1862 by the English naturalist Henry W. Bates, who had spent the years 1849 to 1860 collecting insects, especially butterflies, in the forests of Brazil. Surprisingly few examples of mimicry are known among amphibians. Although dendrobatid frogs are extremely toxic, and one might well expect them to have numerous mimics, only a small number are known. These include the edible *Eleutherodactylus gaigeae* (Leptodactylidae) which is found together with *Phyllobates lugubris* and *P. aurotaenia* (Dendrobatidae), as well as in the intervening territory where an ancestral form of *Phyllobates* probably once lived. Unlike the dendrobatids it resembles, however, the non-toxic *E. gaigeae* is active at night. Nevertheless, during the day, it is sometimes found in the leaf litter where the *Phyllobates* spp. seek shelter when they are pursued. According to Myers and Daly (1983), the different rhythms of activity may therefore be of no significance insofar as potential daytime predators are concerned. There is, as yet, no experimental evidence to support this presumed example of Batesian mimicry, or that of *Lithodytes lineatus* (Leptodactylidae) which is supposed to be a palatable mimic of the slightly toxic *Dendrobates femoralis* (Dendrobatidae).

Other amphibian examples of batesian mimicry are known among North American salamanders. The red phase of *Plethodon cinereus* (Plethodontidae), for instance, is apparently a mimic of the toxic eft stage of *Notophthalmus viridescens* (Salamandridae), while, in parts of its range in the southern Appalachian Mountains, populations of the distasteful black, red-cheeked *Plethodon jordani* have red markings on the sides of their heads which are mimicked by the palatable *Desmognathus imitator* (Plethodontidae). In regions where *P. jordani* has red legs, some individuals of *Desmognathus ochrophaeus* are mimetic (Howard and Brodie 1973). In the Great Smoky Mountains, *P. jordani* is mimicked by *D. imitator* which is considered to be palatable, while, in the Blue Ridge Mountains, *P. hubrichtii* has noxious skin secretions and serves as a model for some individuals of *D. ochrophaeus*. The Californian *Ensatina escholtzii* (Plethodontidae) not only mimics *Taricha torosa* (Salamandridae) in its dorsal coloration, but also closely matches the latter's yellow eye colour and orange venter. The two are sometimes found together in the same retreat (Stebbins and Cohen 1995).

The mimicry of one distasteful model by another distasteful species is known as 'Müllerian mimicry'. (This phenomenon was first described by F. Müller in 1879.) A noxious species only benefits from its warning coloration because some of its members are sacrificed in teaching would-be predators to avoid it. Therefore, if one or more aposematic and unpalatable species mimic one another, the numerical losses incurred in teaching enemies not to attack them are shared and proportionately reduced, to the mutual benefit of each. In the case of Batesian mimicry, on the other hand, the mimics are normally scarce in comparison with their models. If this were not the case, predators would too often encounter palatable mimics and therefore not only would fail to learn to avoid the models but also, on the contrary, might even be encouraged actively to seek them out.

Müllerian mimicry has been recorded among amphibians, but few examples are known. (In the other animal taxa, in which it is common, several different Müllerian mimicry complexes may occur within the same habitat, and by no means all aposematic species share the same warning colour patterns.) The aposematic coloration of the toxic American salamanders *Pseudotriton ruber* and *P. montanus* (Plethodontidae) has been interpreted as a case of Müllerian mimicry with the coloration of the highly toxic *Notopthalmus viridescens*. The effectiveness of the models is probably reinforced by the presence of the toxic mimic, because there is a correlation between size and degree of toxicity in *P. ruber*. This is most toxic when it reaches the size of the efts of *Notophthalmus* spp., and its level of toxicity declines as it gets larger. The potency of the skin toxin and the degree of palatability of aposematic amphibians varies considerably (see Stebbins and Cohen 1995).

Coevolution involving toxins may occur between predators and their prey. This is the case with the garter snake (*Thamnophis sirtalis*) which preys on the toxic newt *Taricha granulosa*. The latter possesses the neurotoxin tetrodotoxin in its skin. Garter snakes exhibit varying degrees of resistance to this, depending upon whether or not they occur together with the newts (Brodie and Brodie 1990).

Venomous New World coral snakes (*Micrurus* spp.) (Elapidae) (Fig. 51) are mimicked by various harmless or mildly poisonous 'false' coral snakes (e.g. *Erythrolamprus, Atractus, Lampropeltis* and *Pseudoboa* spp.: Colubridae). For example, the widespread colubrid *Rhinobothryum lentignosum* of the Amazon basin is strikingly similar to the venomous 'triad' coral snakes of the same area. They all have a pattern consisting of groups of three black rings or triads. The individual rings of the triads are separated by yellow or white rings, while the triads themselves are separated by red rings (Goin et al. 1978). Clearly, both Batesian and Müllerian mimicry are involved.

Coral snakes are so deadly that some investigators have expressed doubts as to whether they could actually serve as models, since potential predators are almost invariably killed, and never therefore learn to avoid their mimics. As an alternative, it has been suggested that rear-fanged, mildly venomous species of snake with coral snake patterns may be the models while nonven-

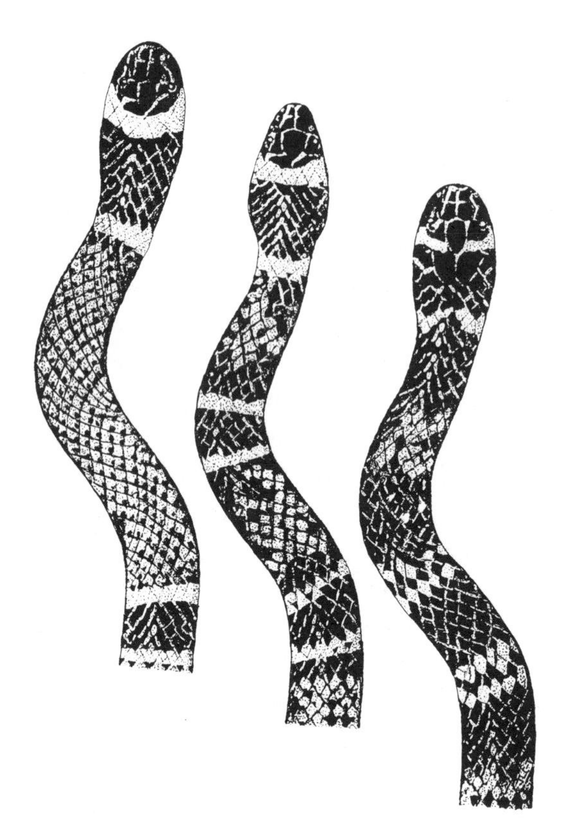

**Fig. 51.** Two venomous coral snakes, *Micrurus diastema* (*left*) and *M. elegans* (*right*), with their harmless colubrid mimic *Pliocerus elapoides (centre)* whose coloration combines the patterns of both models. (Cloudsley-Thompson 1994; after Irish FJ in Greene and McDiarmid 1981)

omous and front-fanged species are respectively their Batesian and Müllerian mimics (Mertens 1966). The name Mertensian mimicry has been applied to this hypothesis (Wickler 1986). Venomous coral snakes are indeed far less numerous than the mildly poisonous colubrids they are thought to mimic. I am somewhat doubtful, however, as to whether any snake could be so poisonous that all animals bitten by it are killed. Moreover, many predators have both natural and acquired immunity against snake and other venoms (see Cloudsley-Thompson 1994). The paradox of the deadly model has also been discussed by Greene and McDiarmid (1981), Pough (1988) and by Roze (1996). Mertensian mimicry is now generally regarded as a version of Müllerian and not as a completely different category of mimicry. The situation is evidently extremely complex, and it is worth mention that scavenging birds, which quickly consume other snakes killed on the roads, apparently leave dead coral snakes undisturbed. So the conspicuous coloration may be related more to distastefulness or an unpleasant smell than to venom!

## 6.4 Structural Defences

Whereas the amphibian skin is thin and not infrequently protected by toxic secretions, the reptilian skin is covered with tough scales composed of keratin. Keratin is the dead material of which hair, feathers, horns, hooves and nails are also composed. In crocodiles and tortoises, the keratin of the skin is continually being rubbed away, but in lizards and snakes, it is sloughed at intervals, often several times a year. In amphibians, the surface (stratum corneum) of the skin is also periodically shed. The skin of salamanders breaks around the mouth and its owner wriggles forward. The front limbs are first pulled free, then the hind legs which are used to push the moulted skin backward on the tail. The clump of old skin is then often eaten by the salamander. In anurans the skin splits down the back, the limbs are pulled free and the cast skin wound towards the mouth and swallowed (Larsen 1976).

Some modern reptiles have soft skins with very thin scales. In crocodiles and certain lizards, however, the dermal parts of the scales contain small plates of bone known as 'osteoderms' or 'osteoscutes' which make the skin exceedingly tough. These plates may be attached to the underlying skull bones of the head region. The chelonian shell consists of a thin outer layer of keratin plates or laminae and an inner layer of bony plates. The laminae do not coincide with the bony plates but overlap them, thereby adding strength to the structure. They increase in size during growth by the addition of new material around and beneath them, but are not a reliable guide to the age of a tortoise.

The girdle-tailed lizards or zonures (Cordylidae) of southern Africa have a remarkable armour of spiny scales which not only make them difficult for predators to swallow, but also enable them to wedge themselves into rock crevices whence it is extremely difficult to extricate them. Spines are also accentuated in the spiny lizards (*Sceloporus* spp.) (Iguanidae) as well as in *Calotes* spp. and in the spiny-tailed lizards (*Uromastyx* spp.) (Agamidae). They have evolved in the two families (three genera) by parallel evolution.

The evolution of the chelonian shell has long been a puzzle. It is now believed, contrary to previous intuition, that dermal armour did not initially function defensively. The ancestral chelonian had no armoured shell: its spine was flexible and the animal was only slightly smaller than most modern turtles. As its descendants became larger, however, they needed stronger bones. Their fossil remains indicate that, as they grew bigger still – up to a metre in length – turtles required even larger vertebrae to support the stronger muscles necessary for swimming by flexing the spine. With increasing size, osteoderms began to cover the animals' backs, and these in turn restricted the ability of the spine to flex so that the limbs had to be used as paddles. Over tens of millions of years, the bones of the spine and the osteoderms apparently fused into a single piece, as did the ribs. The resulting carapace then served to ward off predators. Since then tortoises and turtles have become smaller so that most of them are about the same size as their original ancestor, but their appearance has changed completely (Lee 1996).

The greatest development of armour and spines, however, is seen in the dinosaurs. In particular, the horned dinosaurs (suborder Ceratopsia) a major taxon of the Upper Cretaceous ornithiscians, were characterized by the development of an enormous bony frill from the posterior region of the skull. This extended over the neck region and, in *Torosaurus latus* (Fig. 52), half way along the back. Another major division of the ornithiscians was the Ankylosauria. In these animals, the small osteoderms characteristic of more primitive dinosaurs were developed to form a solid carapace fused to the ribs, vertebrae and pelvic girdle: in *Scolosaurus cutleri* (Fig. 53), the armour comprised a series of bands, each of which carried six large spines. Although the enormous, triangular, bony plates of plated dinosaurs, such as *Stegosaurus armatus* (Fig. 54), probably had a thermoregulatory function, they could well have enhanced the formidable appearance of the animals to predatory dino-

**Fig. 52.** Reconstruction of a ceratopsian dinosaur (*Torosaurus latus*) (length ca. 8.5 m), Upper Cretaceous. (Cloudsley-Thompson 1994)

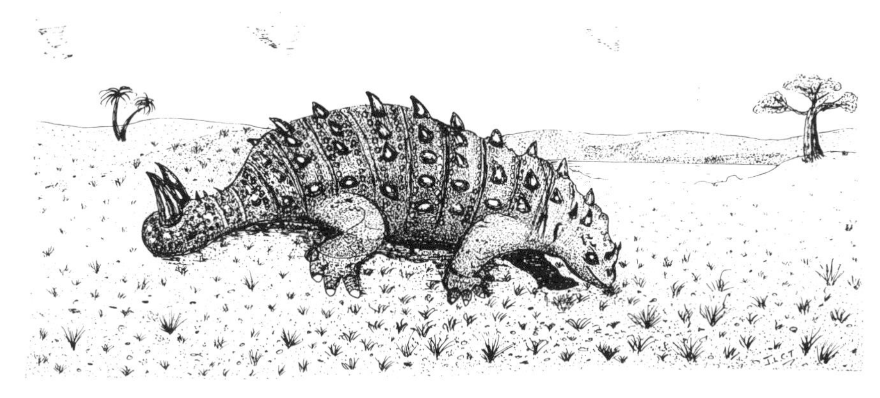

**Fig. 53.** Reconstruction of an ankylosaurian dinosaur (*Scolosaurus cutleri*) (length ca. 6 m), Upper Cretaceous. (Cloudsley-Thompson 1994)

**Fig. 54.** Reconstruction of a plated dinosaur (*Stegosaurus armatus*) (length ca. 6 m), Upper Jurassic. (Cloudsley-Thompson 1994)

saurs. Nothing comparable, even on a much smaller scale, is to be found in recent reptiles.

## 6.5 Secondary Anti-predator Devices

Secondary anti-predator devices are invoked only in the presence of an enemy. The simplest response of most prey animals is to fly or run away, while burrowing forms disappear as fast as they can into their shelters or retreats. This may not always be possible, however, and self-defence with whatever weapons are available may then become necessary. Some animals, including tortoises, are so well endowed with physical defences that they stand their ground and make no attempt whatever to escape. Others, equipped with formidable offensive weapons - especially teeth - readily employ them for defence. The same would have been true of large extinct amphibians, dinosaurs and so on. While cold and sluggish, some agamid lizards will stand and fight. When warm, however, they immediately resort to flight (Hertz et al. 1982). In the case of *Uromastyx microlepis* the change in behavioural response is accompanied by a change in colour. When very cold, *U. microlepis* is dark grey and shows a death-feigning response. On warming, however, the animal first defends itself - exhaling with loud hisses and swinging its scaly tail, a most effective weapon. Only when considerably warmer does it recover its natural cryptic coloration and run away (Cloudsley-Thompson 1991).

Lizards that forage widely and rely upon rapid flight to escape predation are usually streamlined and have a low ratio of clutch to body weight. A large clutch of eggs would hamper such species in running, and thus reduce their ability to escape: it would also reduce their foraging efficiency. Cryptic, sit-and-wait foragers, on the other hand, tend to be short and robust, and to produce relatively large clutch masses. Since they do not rely upon speed, either for foraging or for escaping, they can afford to bear heavy clutches. Thus, relative clutch mass, body shape, and foraging and escape tactics are all interrelated in lizards (Cloudsley-Thompson 1994). Two widespread specialized locomotory escape mechanisms of lizards are running on soft sand, which is often facilitated by toe fringes as discussed in Chapter 4 (Luke 1986), and climbing relatively smooth, vertical surfaces with the aid of highly evolved morphological adaptations of the digits. These adaptations are found in geckos, some iguanids, and a few species of skinks.

Venomous reptiles, such as vipers and cobras are usually secretive and cryptic in appearance, but, when they are threatened and cannot escape, indulge in threatening aposematic displays. The Indian cobra (*Naja naja*), for instance, rears up with its hood spread and showing the characteristic spectacle markings (Fig. 55) which look like enormous eyes, many arboreal snakes inflate their throats, and rattlesnakes throw their necks and heads into S-shaped coils. In most cases, a snake strikes rapidly in the darkness and then disappears before it can be identified. The characteristic punctures made by the fangs, however, often enable diagnosis to be made: the presence of two bleeding fang marks is a clear indication that the bite is that of a dangerously poisonous species. If other tooth marks are also present, a cobra is responsible; if not, the bite may have been inflicted by a viper.

Many amphibians and reptiles, when first disturbed or attacked, become motionless. By becoming immobile, they may escape from their predicament because many predatory animals are not stimulated to attack dead prey, while others do not always eat their victims immediately after killing them. The corpses are stored to eat later, so the amphibian or reptile feigning death may be left alone and consequently escape. Death feigning or 'thanatosis' may also delay recognition of the prey for what it is or reduce the attentiveness of the predator. Some snakes roll themselves into tight, motionless balls which protects their vulnerable heads when they are attacked. Hognose snakes (*Heterodon nasicus*: Colubridae) sometimes strike themselves sufficiently severely to cause bleeding, while the African savanna monitor (*Varanus exanthematicus*) adopts a rigid posture when it feigns death (see Cloudsley-Thompson 1994).

**Fig. 55.** Indian cobra (*Naja naja*) with hood expanded (length ca. 1.8 m). (Cloudsley-Thompson 1980)

Certain amphibians and most reptiles, whether venomous or not, bite in defence, often with a sideways lunge of the head. Salamanders have been recorded as biting both snakes and shrews when attacked by them. Some amphibians also vocalize: the distress calls of certain frogs have been aptly described as screams. These may well startle and deter aggressors, while the explosive bark of the North American grey tree frog (*Hyla versicolor*) has been shown to repel shrews (Stebbins and Cohen 1995).

## 6.6 Autotomy

The tuatara (*Sphenodon punctatus*), many lizards, some amphisbaenians and a few snakes are able to shed their tails when attacked by a predator. Some salamanders do the same. The efficiency of this as an escape mechanism is apparent to everyone who has tried to catch lizards by hand in the field. Broken tails have often been removed from the stomachs of predators and , in captivity, lizards will often eat their own tails after these have been autotomized. The violent wriggling of the tail fragment may continue for several minutes while its owner makes its escape. In some species the bright colour of the tail contrasts with that of the body, and probably adds to the distraction it creates.

Caudal autotomy is an adaptation which involves sacrifice that a reptile cannot really afford, but which is preferable to the alternative of almost certain death. Its costs include: (1) energy lost in replacing the tail must be diverted from reproduction, growth and other functions; (2) a potential loss in social status; (3) the tail cannot be used again as an anti-predator mechanism until it has been regenerated; and (4) lizards cannot run so fast and their chances of escape are reduced after the tail has been shed (Punzo 1982). In compensation, tailless lizards show a reduction in activity and flight distance: they adopt a cryptic anti-predator strategy. The subject of caudal autotomy is discussed in Cloudsley-Thompson (1994) and has been thoroughly reviewed by Arnold (1984, 1988) and by Bellairs and Bryant (1985).

## 6.7 Startling Behaviour

Startling or 'deimatic' behaviour (from the Greek word meaning to frighten) is a common response to predatory threats, and its function is to disconcert an aggressor and persuade it to go away. I have already mentioned the jumping of Pygopodidae: this probably has a deimatic function. The bright colours of aposematic amphibians are usually concealed – unless the animals are threatened – when their sudden appearance may be another instance of deimatic behaviour. It could, however, also be a genuine warning that the potential prey is toxic or might harm its enemy in some way. On the other hand, deimatic displays are often sheer bluff; there is no clear distinction between the two. An amphibian of reptile may be unpalatable to one predator but not to another, so that its deimatic display is a warning in the first instance and bluff in the second. Mimicry may also be involved, both Batesian and Müllerian.

A sudden hiss is a startling sound at any time, but when a grass snake (*Natrix natrix*) (Colubridae) hisses, is it mimicking an adder (*Vipera berus*) (Viperidae) or is the hiss alone sufficient to deter the enemy without an element of mimicry being involved? Of all the sounds emitted by reptiles, the hiss is the most characteristic. When disturbed, turtles, tortoises, crocodiles, alligators, monitor lizards, chameleons and snakes, both venomous and non-poisonous, make hissing sounds. Many animals other than reptiles also hiss in self-defence (Cloudsley-Thompson 1994).

Warning sounds include the stridulations of saw-scaled vipers (*Echis* spp.), horned vipers (*Cerastes cerastes*) and the sand viper (*C. vipera*) of North Africa. In these, the orientation of the lateral scales of the trunk has changed so that their keratinous keels are inclined with the posterior edges downwards. The snakes can therefore stridulate by rubbing these scales against each other and without losing respiratory water as they would by hissing. The acoustic characteristics of the rattling sounds of rattlesnakes (*Crotalus* spp.), which likewise do not result in loss of water, have been analysed by Fenton and Licht (1990) who found that they all conform to the same general pattern. They are medium-intensity, broad-based sounds with rapid onset and no structured change in frequency pattern over time. Therefore they must be a signal designed to attract the attention of other animals, whose responses will reflect their hearing characteristics and previous experience. While the majority of snakes are silent, there are a few that hiss in the normal way and others that vibrate their tails against vegetation. Sweet (1985) suggested that the tail vibrations of non-venomous gopher snakes (*Pituophis melanoleucus*) producing an audible buzz in dry leaves may be mimicking the sounds of rattlesnakes (*Crotalus viridis*). In several species of *Pituophis* a development of the thorax produces a particularly loud hiss (see Cloudsley-Thompson 1996).

One form of deimatic behaviour is an apparent increase in size, often achieved by inflating the lungs and turning the body so that the broadest part is exposed to the enemy. The defence reaction of the common toad (*Bufo bufo*) consists of adopting a peculiar posture in which the animal puffs out its body like the frog in the fable who lost his life by attempting to reach the dimensions of an ox! By inflating its lungs, the toad can increase its size by as much as a half. At the same time, the bulky body is raised from the ground and tilted towards the side from which danger threatens. Deimatic inflation of the body is not always bluff, as we have seen. In the present instance, its principal function is to prevent the toad from being swallowed by a snake. Venomous snakes also increase their size during threat displays. Cobras spread their hoods; the rear-fanged African boomslang (*Dispholidus typus*) (Colubridae) inflates both trachea and lung to such an extent that the snake resembles an enormous sausage. Bird or twig snakes (Fig. 29) are timid animals, relying on their intricate coloration to escape notice. If pressed, however, they resort to a display similar to that of the boomslang and flatten their heads like an ace of spades. Although they seldom bite in defence unless provoked, they possess a potent venom that has resulted in human fatalities. The relatively enormous frill of *Chlamydosaurus kingii* (Fig. 22) is used primarily in interspecific competition, and especially for territorial displays by males. It is, however, also employed to deter potential predators.

The tongue and lining of the mouth are often brightly coloured in lizards and snakes, and this internal coloration is associated with the habit of opening the jaws widely when threatened. For example, not only do vine snakes produce warning sounds by vibrating their tails as terrestrial snakes also do when they rattle their tails among dry leaves, but furthermore the blue-black lining of the open mouth of *Oxybelis* spp. serves as a threat display. In addition, the front part of the body of *Dasypeltis* spp. may be inflated as a threat. Among poisonous snakes, deimatic exposure of mouth colours is a threat, but in harmless species it is just bluff.

The violent writhing of an autotomized tail may also startle an enemy as well as divert its attention so that the prey can escape. The defensive responses of Amphibia are discussed by Duellman and Trueb (1986), Stebbins and Cohen (1995) and Zug (1993) among others, while defensive strategies among reptiles have been considered in further detail in Cloudsley-Thompson (1994) as well as by various authors in Gans and Huey (1988). I will not, therefore, discuss this fascinating subject further, except to mention that some reptiles, such as the European grass snake (*Natrix natrix*), can produce unpleasant smells that are difficult to wash off. Always point chelonians away from yourself when you hold them because their urine is used in defence and stinks! After burying their eggs, female tortoises often urinate on the soil so that would be predators are warmed off.

# 7 Nutritional Diversity

Insofar as their diet is concerned, amphibians and reptiles have much in common. Most are carnivorous and, although many amphibians are herbivorous in their aquatic larval stages, they become carnivorous after metamorphosis. Immediately after hatching, both amphibians and reptiles receive nourishment from the remains of the yolk formerly present in the egg. In the case of tadpoles, the mouth is a mere dimple and is not connected with the gut. A glandular area on the skin of their heads, which secretes sticky material, enables some very young tadpoles to spend much of their time hanging from aquatic plants. As their yolk is used up, however, the adhesive organs are lost and tadpoles assume a more typical shape with ovoid heads and bodies and longer tails. Their mouths develop and become united with their alimentary canals, while in the Anura, gill clefts are formed through which the inside of the mouth is connected with the water in which the tadpole is living. Water is taken in through the mouth and expelled through the gill clefts (Frazer 1973).

Horny lips appear on the jaws, as well as rows of small, tooth-like structures whose number and appearance vary from one species to another. The latter are used for rasping off portions of vegetation, algae and diatoms. As tadpoles grow, their feeding habits change. Developing limbs require a source of animal protein, and the little creatures soon become carnivorous, feeding on corpses and even eating other tadpoles. As with most generalist feeders, diet depends upon the availability of suitable food. To quote a German proverb: 'In der Not frisst der Teutfel Fliegen' (During famine, the devil sups on flies).

## 7.1 Herbivory

Terrestrial vertebrates were widespread in the Devonian period and a number of different forms are known, but fossils are somewhat scarce. Although the

Permian – Triassic extinction severely reduced the diversity of terrestrial vertebrates, floral turnover was rapid, and, during the Early and Middle Triassic, herbivore faunas were dominated by generalized browsers. These relied on plant productivity within about 1 m of the ground, implying that herbaceous pteridophytes (ferns, club-mosses and their allies), small seed-ferns and young plants of all types, including primitive pseudocarpaceous conifers, cycadeoideans and ginkgoes, would have received the brunt of browsing activity. An increase in browsing height from 1 to 4 m occurred with the advent of the prosauropod dinosaurs in the Late Triassic, and reflected an evolutionary change of great ecological importance.

Not only is herbivory characteristic of anuran larvae, but also it may occur in other amphibians. For instance, aquatic salamanders of the genus *Siren* have been reported with large quantities of *Elodea* sp. and other vegetable matter in their alimentary canals; and *Bufo marinus* may eat vegetable scraps, both fresh and rotting (Duellman and Trueb 1986). Brazilian tree frogs (*Hyla truncata*) include fruit in their diet, and the same is probably true of many other species of Anura.

Herbivory has not been well developed among amphibians. In contrast, as already mentioned, plants have been exploited by reptiles from earliest times. Many of the dinosaurs were strictly herbivorous. The Mesozoic era was one of climatic uniformity: temperatures were tropical or sub-tropical over much of the surface of the earth, and vegetation was abundant. Even giant brontosaurs with their relatively tiny jaws and weak teeth, as well as armoured dinosaurs with very small teeth, were able to crop enough vegetation to sustain their vast bodies under the equable conditions of the time, although they probably grew rather slowly and enjoyed very long lives.

During the Triassic, six major methods of food processing evolved among herbivorous vertebrates (Wing and Sues 1992). In the Late Triassic, however, there was a general shift from oral to gastric mill processing of the vegetation (see below). This must reflect an interaction between the herbivorous reptiles and the plants upon which they browsed. By the Middle Jurassic, dinosaur-dominated herbivore faunas were well developed. The largest sauropods could have browsed at a height of 10 m in a quadrupedal posture, and very much higher when they reared up on their hind legs, propping themselves with their tails. Their dentition consisted of peg-like teeth showing few signs of wear. This implies that the teeth were used in cropping and the food broken down in a gastric mill: the capacious abdominal cavity suggests the presence of a voluminous gut.

The second most abundant terrestrial vertebrates at this time were the stegosaurs (suborder Stegosauria) (Fig. 54). These heavily armoured reptiles reached a length of up to 7 m and a weight of up to 5 tonnes. They had narrow, elongated snouts with simple, spatulate teeth and fed mainly within a metre of the ground unless they reared up on their hind legs. Other herbivorous dinosaurs of the period included ornithischians such as *Camptosaurus*. These developed a number of remarkable specializations for herbivory which were unique among reptiles. Their teeth show evidence of wear, which proves that they must have been capable of grinding or chewing their food. Study of their skulls indicates that they possessed muscular cheeks, as do mammals, and they even developed a secondary palate so that their food and air passages were separate. This enabled them to retain food in the mouth, to chew it and breathe at the same time. As Halstead and Halstead (1981) remarked, 'This condition parallels the situation in the mammals to a remarkable degree and is one of the most dramatic examples in the fossil record of convergent evolution – two entirely separate lines evolving similar structures to subserve similar functions' (p. 25).

Like the Camptosauridae, the Dryosauridae and Hypselophodontidae had well-developed capabilities for chewing their food. They foraged mainly within 1–2 m, and perhaps up to 4 m, from the surface of the ground. They were very abundant in many communities. Even smaller herbivorous and omnivorous Jurassic reptiles included Sphenodontidae, which were relatively ubiquitous, and lizards (Wing and Sues 1992). Most of these would have foraged close of the ground, but at least some were arboreal.

Large, high-browsing sauropods dominated the Jurassic terrestrial herbivore fauna. Although their metabolic rates would have been relatively low, they would nevertheless have required large amounts of food. Yet the major groups of Jurassic trees do not appear to have produced large quantities of foliage, nor is there evidence that they had high growth rates. Furthermore, many living conifers and cycads have toxic foliage, spiny leaves and a tendency to retain dead foliage around the trunk. These may have been important defences for their Jurassic ancestors. The thick texture of many Jurassic trees may also have been a deterrent to herbivores as well as an adaptation to dry climates. Paradoxically, in the light of the high-browsing adaptations of sauropods, the plant groups most likely to have provided abundant fodder were short shrubs and herbs! The conflict between animal and plant evidence suggests that, although the diversity of large herbivores was high, their numbers and biomass per unit area would have been relatively small.

By the Early Cretaceous period there was a marked difference between the tetrapod faunas of the northern and southern hemisphere, although some interchange between the two assemblages still took place. In the north, there was a decrease in the diversity and abundance of high-browsing sauropods, while in the south these continued to be the dominant element among the herbivores. Stegosaurs became much less important while the ankylosaurs increased in diversity and abundance. The latter were present in some Jurassic faunas, but did not diversify until the Cretaceous. The Ornithopoda also radiated during the Cretaceous: some early forms, such as *Iguanodon* and *Tenontosaurus*, reached a length of 10 m and a weight of up to 2 tonnes, so they would have been able to browse at higher levels above the ground than their Jurassic predecessors. (The first ceratopsians also appeared in the Cretaceous.) In contrast, sauropods dominated in the south, where hadrosaurid ornithopods were quite rare.

Both Hadrosauridae and Ceratopsidae had elaborate dentitions with closely packed, interlocking teeth. The increase in abundance and diversity of angiosperms coincided with an increase in the abundance and diversity of these two groups of large herbivores. Both were adapted to extensive chewing of plant food, and probably foraged in herds. The coincidence of these two radiations may well reflect diffuse coevolution, in which the increased productivity of angiosperms led to the success of these groups of herbivores. In the southern hemisphere, angiosperms were a less important element of the vegetation, and Titanosauridae, rather than low-browsing ceratopsians and hadrosaurs, were the dominant herbivores. After the extinction of the dinosaurs at the end of the Mesozoic era, mammals became the most important terrestrial herbivores (Wing and Sues 1992).

Of the four major taxa of extant Reptilia, two are entirely carnivorous and two, as we have already seen, eat both plants and animals. Crocodilians live entirely on flesh and one subfamily (Gavialinae) comprising the Indian gharial (*Gavialis gangeticus*) feeds solely upon fish. Crocodilians are not capable of swallowing large prey whole and, consequently, have to tear off portions of flesh by a series of sudden jerks and rapid twists of the body. Most crocodiles and alligators are insectivorous when they hatch, graduating to fishes and, finally, to mammals as they grow large (Cott 1961). Snakes, too, are entirely carnivorous, but Chelonia and lizards may be either herbivorous, omnivorous or carnivorous. Incidentally, the generic name *Gavialis* originates from a printer's misreading of *Garialis* which, of course, comes from the Indian name 'gharial'. Once the name had been published, albeit incorrectly, it could not be changed unless this was approved by a resolution of the International Commission on Nomenclature – and no such resolution has been proposed.

Relatively few modern reptiles are, nevertheless, entirely herbivorous, apart from some turtles, tortoises, and some lizards. The green sea turtle is one of the few extant specialist herbivores among reptiles. Most reptiles are generalists, and include foliage, fruits and flowers in their diet. As already mentioned, the majority of green turtles eat 'turtle grasses', but some populations feed on seaweeds (algae). No individual green turtle can eat both, because different intestinal floras are required to break down these structurally dissimilar plants. Most reptilian herbivores show a preference for young, growing foliage. This is more easily digested than older vegetation and has a higher content of protein. Green turtles normally graze on already grazed plots and continually crop the new growth. When introduced to new areas, they bite the sea grass close to its base, allowing the older fragments to float away and thus establish a large grazed pasture that can be harvested daily (see Zug 1993). Chelonians are unusual among reptiles in being toothless. Instead of having teeth, the jaws are provided with horny beaks shaped to fit over the underlying bones.

Most land tortoises (Testudinidae) are mainly herbivorous, feeding somewhat indiscriminately on whatever vegetation and fruits are available, while plants comprise more than 75% of the diet of terrapins (Emydidae). These herbivorous chelonians include both large and small species, as well as semi-terrestrial and aquatic forms (e.g. *Cuora*, *Hardella*, *Ocadia* and *Pseudemys* spp.). Other terrapins, including *Chrysemys* spp. are facultative herbivores and can survive and grow on a vegetarian diet when animal prey is not available. Most predominantly herbivorous reptiles will occasionally eat animal matter when they come across it.

A number of lizard species are generalist herbivores. These include many Iguanidae, a few Agamidae (e.g. *Uromastyx* spp.) and some skinks (Scincidae) (e.g. *Corucia* spp.). Many of these are large, suggesting that the capture of small prey may be energetically expensive in regions where prey is scarce. Many iguanids have been found to be herbivorous from birth.

Other lizards may become more herbivorous as they grow older. The rainbow lizard (*Agama agama*) is primarily insectivorous at Ibadan, Nigeria, feeding almost entirely on ants. Larger males are, however, cannibalistic. Only occasionally is vegetation eaten there, and then chiefly by adult males which show a preference for coloured objects (Harris 1964). At Nsukka, eastern Nigeria, on the other hand, a large amount of vegetable matter is eaten by adults during the dry season, which may reflect the scarcity of insect food available at that time of year (Cloudsley-Thompson 1981). Zug (1993) tabulates a number of reptilian herbivores whose diets, when adult, consist predominantly of plant matter. Some of them possess a cellulolytic microflora that is able to break down cellulose in their digestive tracts and/or in colic modifications of the hindgut.

Herbivory in reptiles was discussed by Pough (1973) and is the subject of a recent book by King (1996), who addressed the question as to why there should be so few herbivorous reptiles in the world today in comparison either with mammals or with the impressive reptilian radiations of past eras. Mesozoic herbivores enjoyed two major periods of increasing diversity, Triassic–Early Jurassic and Middle–Late Cretaceous. During the first of these, changes in the flora of gymnosperm plants may have controlled the adaptations of the herbivores through changes in food quality, such as toughness, with which the earlier therapsids were not adapted to cope. The early dinosaurs were probably more successful as a result of better gut processing. Most of the reptilian herbivore taxa of this period had simple or reduced dentition. In contrast, the Cretaceous increase in reptilian diversity was marked by a corresponding diversity of tooth form and jaw mechanisms. This might well have been the result of coevolution between the herbivorous dinosaurs of the period and the fruiting structures of the later Mesozoic angiosperms. These became well defended by hairs, thorns, hard outer coverings and so on, which protected them from being eaten before the seeds had matured. No doubt herbivorous reptiles would also have played an important part in seed dispersal.

The giant sauropod dinosaurs of the Mesozoic used gastroliths or stones that they had swallowed to grind vegetation in muscular gizzards. Galapagos tortoises and lizards (*Sauromalus* spp.) swallow sand and gravel probably for the same reason. Two main theories have been proposed to explain lithophagy: either to produce ballast in aquatic forms, or to act as grinding stones to break down food. In crocodiles, it seems reasonable to assume that gastroliths have the primary function of ballast. When items such as eggshell and cuttlebone are ingested, they may provide an extra source of calcium prior to egg-laying.

A considerable amount of information is, of course, available about feeding in living reptiles, not only from how teeth and jaws work but also from the structure and physiology of the alimentary canal. A few examples of the mummified gut contents of dinosaurs are also known. Perhaps the most famous is that of the hadrosaur *Apatosaurus* whose presumed stomach contents comprised conifer needles, twigs, seeds and other parts of land plants. This was one of the early pieces of evidence which showed the duck-billed hadrosaurs to have fed at least partially on land plants and not on soft water vegetation as previously supposed. In addition, a coprolith attributed to an ornithopod dinosaur apparently contained bits of fibrous undigested plant matter. Finally, footprints have been preserved with plant matter in such a way that they seem to have been made by dinosaurs browsing on the foliage of both conifers and of angiosperm trees.

Fossilized parallel trackways of dinosaur footprints also provide evidence of herding, as do the mass accumulations of skeletons of dinosaurs such as *Iguanodon* and *Plateosaurus* spp. All that these large numbers of skeletons really indicate, however, is that the dinosaurs were common and successful in life. We cannot actually prove that they had any specific association with one another while they were alive. Nevertheless, in the case of *Iguanodon*, parallel trackways are known as well, which would indicate some kind of herding. King (1996) concluded that herbivore-plant interactions were probably responsible for changing rates of specialization and extinction in dinosaurs.

## 7.2 Carnivory and Omnivory

The evolution of the pelycosaurs and early therapsids took place in two types of trophic systems. In one, the food base was provided primarily by aquatic plants, while in the other, it lay in terrestrial plants. The second type appears to have diverged from the first at some time during the Carboniferous period, but this may have occurred even earlier. *Dimetrodon* was a top predator feeding on aquatic and semi-aquatic animals, probably including species of *Eryops*, *Diadectes* and *Ophiacodon*. Since it was the most adundant of the large reptiles, it may also have preyed on fishes, amphibians and on reptiles smaller than itself, while, during the later part of its existence, it was almost the only large animal in the ecosystem and must therefore have fed entirely on fishes, amphibians, and reptiles such as *Captorhinus*. Adaptation for feeding on larger animals, evident in the mechanism of the jaws, undoubtedly evolved during the earlier parts of the history of *Dimetrodon*, and was retained with little modification when the faunistic composition of the community changed (Olson 1986).

In all the stages of their life cycles, extant urodeles and caecilians, as well as anurans after metamorphosis, tend to be opportunistic carnivores preying on any animal that is small enough to be captured and swallowed (Larsen 1992).

Although it often appears that amphibians show little discrimination, they are in fact not wholly lacking in food preferences. Jaeger and Rubin (1982) designed an ingenious experiment to determine whether the red-backed salamander (*Plethodon cinereus*) determines the caloric profitability of potential prey (flies of different size) by innate recognition of their size, or by learning through previous experience with various prey types. They concluded that the animal does not rely on an inherited ability to assess profitability but that it learns, through foraging experience, to balance the gross calorific value of the prey – which can be estimated by size – against the rate at which it can be assimilated once it has been ingested. This is a variable depending, in the case of arthropods, on their chitinous content. Again, the Californian slender salamander (*Batrachoseps attenuatus*) seems selectively to ignore smaller prey and to prefer springtails to oribatid mites of similar size, perhaps because the latter have very thick exoskeletons. Factors such as differing nutritional needs between growing young and adults, sex-related requirements (e.g. courtship, vocalization, defence of territory, production of yolk for eggs) and seasonal shifts in the composition and abundance of prey need to be taken into account when studying foraging behaviour (Stebbins and Cohen 1995).

Most salamander larvae feed indiscriminately on aquatic invertebrates of appropriate size, but some larval newts feed only on algae as already mentioned. It seems probable that developmental changes in food and foraging strategies are partly associated with the extent to which the jaws can gape, a matter discussed in some detail by Duellman and Trueb (1986). Smaller amphibians are limited to small prey, but, within size classes, some species have much smaller gapes than others. Those with larger gapes – leptodactylid and ranid frogs – have a more diversified diet, including relatively large as well as small prey.

The generalized feeding mechanisms of some terrestrial salamanders and primitive frogs involve only slight protrusion of the tongue, which is subsequently used to manipulate the prey against the vomerine teeth whilst the jaws are in use (Fig. 56 above). (Vomerine teeth are attached to the bones of the palate which form the roof of the mouth.) Teeth are shed and replaced throughout the lives of amphibians. In the three extant orders, they are jointed or hinged. Their bases or pedicels are attached to the jaws, while the crowns, in turn, are loosely attached to the bases. Amphibian teeth are typically simple, conical, two cusped (or bicuspid) structures usually recurved in the direction of the rear of the buccal cavity. Caecilians and a few frogs have unicuspid teeth arranged in one or two rows on all the jaw bones. Most frogs lack teeth on the lower jaw, however, while a few have none on the upper jaw. Toads (*Bufo* spp.) have no teeth at all. Depending upon the species, the teeth are used for holding, crushing, shredding, or manoeuvring the prey down the throat. There is little or no mastication.

In most salamanders, the tongue is thrust forward by contractions of the muscles associated with the hyoid apparatus to which the tongue is attached. (The hyoid bones evolved from the second gill arch of the ancestral fish-like vertebrate, and support the floor of the mouth in amphibians.) Some salamanders have the tongue attached anteriorly to the floor of the mouth although it can be protruded from behind (Fig. 56, above), while in others it is free and attached to an extensible base (Fig. 56, below). This is the situation in lungless salamanders of the family Plethodontidae, the highly projectile tongues of which can, in many species, be protruded for over a third of the snout–vent length. This takes place so rapidly as to be nearly undetectable to the human eye, and is especially effective in capturing small and highly mobile prey (Duellman and Trueb 1986; Stebbins and Cohen 1995).

The feeding of most advanced anurans consists of a flip of the fleshy tongue, the projectile part of which is separated from the hyoid apparatus. Different families show great diversity of arrangements of the tongue and lingual muscles, but the general principles involved in the capture of prey are

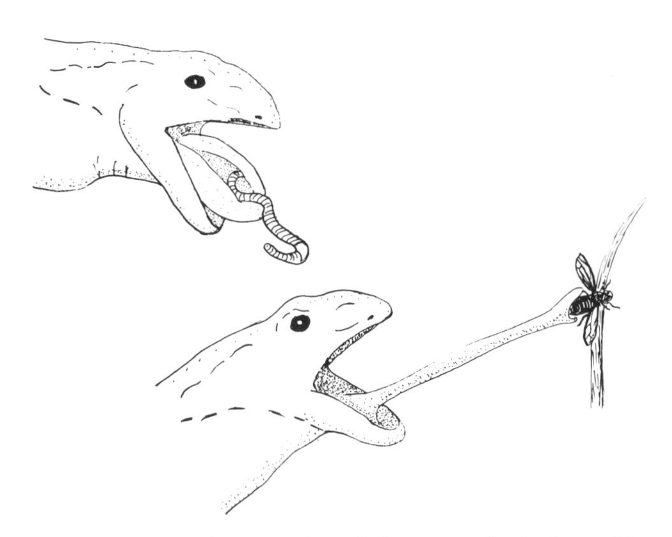

**Fig. 56.** Prey capture by salamanders. Slight protrusion of the tongue in Ambystomidae (*above*); projected tongue of Plethodontidae (*below*)

similar. Two conflicting hypotheses have been proposed to explain how the toad flips its tongue forwards to capture prey: (1) the hyoid bone moves suddenly forward and transfers its momentum to propel the tongue, or (2) intrinsic tongue muscles are stiffened, rotate over the symphysis of the mandibles and catapult the soft tissues forwards. By means of high-speed cinematography and synchronized electromyography, Gans and Gorniak (1982) showed that the tongue is rolled over the symphysis of the mandibles by a complex mechanism of rods formed from stiffened tongue muscles (Fig. 57). As this occurs even when the hyoid is immobilized, the tongue is clearly not flipped forward by the momentum of the hyoid. In both salamanders and chameleons, the protrusible tongue is mounted on the hyoid whose movement everts it or positions it in the mouth for further propulsion. Anurans, however, lack such an intrinsic lingual structure. Indeed, there are two sets of muscle fibres locked into a framework of connective tissue. These are transformed into rods at right angles to their line of action. Swelling of the muscles, therefore, as much as shortening the tongne, rotates the lingual frame and produces a muscular ballista.

Amphibian tongues are usually kept sticky by mucous glands in the pad. Aquatic salamaders, and frogs of the family Pipidae which are completely aquatic, are tongueless. They feed by rapidly opening the mouth and expanding the throat, thereby sucking in both food and water. The water is expelled

**Fig. 57.** Tongue flipping feeding mechanism of an anuran. (After Duellman and Trub 1986, adapted from Gans and Gorniak 1982)

before the mouth is completely closed. This is an effective means of feeding on zooplankton. Many salamanders and anurans depress their eyes whilst swallowing. Since the eyeballs are firm, their lower surfaces, which extend into the mouth, help to force food down the throat. Some Anura also use their fore limbs to cram food into their mouths, and many wipe their mouth and eyes with their fore feet after feeding.

Poison frogs (Dendrobatidae), common inhabitants of rainforest leaf litter in the New World, show an interesting correlation between diet and the toxic compounds in their skins. Species of the genus *Dendrobates*, which are both aposematic and poisonous, have been shown to have diets composed of 50–73% ants, whereas the proportion of ants eaten by non-toxic, cryptic species of the genus *Colostethus* was found to be only 12–16%. The toxic compounds in the skin are lipophilic alkaloids. These are present in a high percentage of the Formicidae eaten by toxic frogs. The latter devour large numbers of ants, composed of relatively few species, whilst cryptic, non-toxic frogs eat smaller numbers of ants but the niche breadths of their diets are high. They include a broader range of prey in their diets than aposematic frogs do. Thus diet, and the subsequent evolution of systems for the uptake of alkaloids, may be the primary character that led to the development of toxic skin. This, in turn, permitted aposematism to evolve, thus leading to the radiation of poisonous species (Caldwell 1996).

Turning now to extant reptiles, two basic types of ingestion are found in predatory lizards. If the prey is small, it is seized between the jaws and killed by being crushed and, in the case of arthropods, having its integument perforated. If, however, the prey is large in relation to the head of the lizard, it is lifted from the ground and killed by crushing or violent shaking. Some species of *Varanus* are strong enough to break the back of a small vertebrate in this way, or at least to rupture its visceral cavity. Arthropods are usually chewed and crushed before being swallowed; larger prey are ingested by inertial feeding. In this, the jaws are suddenly relaxed, and the head of the lizard then slides forward and laterally over the prey which is bitten in another portion of its body. Inertia prevents it from moving when the jaws of the predator shift forward. Lateral movements of the jaw are irregular when the prey is unevenly shaped, and they are sometimes omitted altogether. The animals' tongues are not used in the manipulation of food.

Just as in amphibians, the tongues of some lizards have become adapted for rapid projection and the capture of prey. The rainbow lizard (*Agama agama*) and the moloch or thorny devil (*Moloch horridus*), for instance, can protrude their tongues for a short distance to capture ants and other small insects. These adhere to their sticky tips. This tendency reaches an extreme in chameleons (Chamaeleonidae) whose tongues can be projected to a distance about one and a half times the total length of the body, excluding the tail (Bellairs 1969). When retracted, the chameleon's tongue is in the form of a tube whose walls consist of longitudinal fibres of hypoglossal muscle. These keep the tongue packed into tight pleats on a pointed bone or cartilage at the back of the mouth. The terminal, club-shaped portion of the tongue contains a powerful accelerator muscle which begins where the longitudinal muscle fibres end. The muscle runs radially in a transverse plane, instead of being circular like the muscles of a sphincter. When contracted, the pressure engendered by the accelerator muscle is distributed hydraulically. As they stealthily approach their prey, chameleons hold their heads forward, open their mouths and move their tongues to the front of the jaw. Contraction of the accelerator muscle moves the tongue off the central cartilage and, when the longitudinal muscle is relaxed, the tongue shoots forward like a released spring (Fig. 58). In addition to being glandular and sticky, the tip of the chameleon's tongue has a small depression, the edges of which can be used to grasp the prey. Locusts and other insects that are too large and powerful to be captured by adhesion

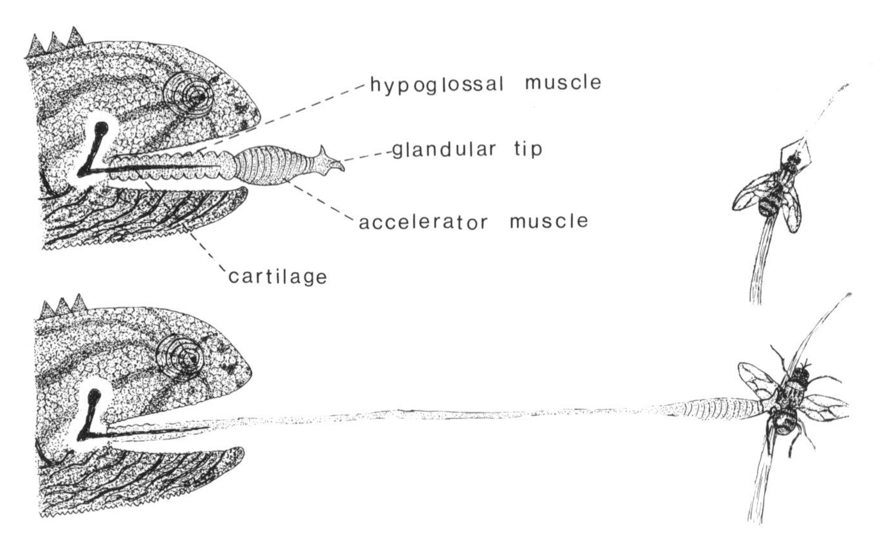


**Fig. 58.** Mechanism of extension of the chameleon's tongue. Tongue packed in tight pleats on a cartilage in the back of the mouth (*above*); tongue projected forward by relaxation of the hypoglossal muscle and contraction of the accelerator muscle (*below*). The prey is trapped by the sticky glandular tip. (Cloudsley-Thompson 1996)

alone are caught by larger chameleons in this way. The tongue is retracted more slowly than it is projected, especially when the prey is heavy.

## 7.3 Snakes

If the Mesozoic era was 'the age of reptiles', then the Neogene (Miocene and Pliocene) surely qualifies for the title 'the age of snakes'! Today, some 80% of lizards weigh less than 20 g as adults and are insectivorous, while 75% of snake species weigh over 20 g when adult and prey mainly on small mammals. With their attenuated bodies they are able to glide down narrow burrows that are quite impassable to most predatory mammals. In this way, competition between lizards, snakes and predatory mammals is largely avoided. However, legless locomotion and small mouths impose difficulties, and the success of snakes in general reflects their diverse solutions to these problems (see below). Some have specialized on eating larger mammals, others on much smaller prey. Specializations are also manifest in the detection, capture, subduing, swallowing and digesting of prey. Various snakes have become adapted to a wide variety of food items ranging from insects to hard-shelled eggs, and from snails to small antelopes. Parallelism and convergence have taken place in many groups. In general, however, Viperidae have proportionately stouter bodies, larger heads and longer jaws than other snakes. They can, therefore, swallow larger prey with fewer jaw movements than can Boidae or Colubridae. Their proteolytic venom also aids digestion (Pough 1983).

The sensory pits of rattlesnakes and other Crotalinae are very sensitive to temperature changes and can detect mammalian prey at night by the warmth of its body, while the paired organs of Jacobson enable lizards and snakes to smell chemicals picked up by the protrusible tongue (see Bellairs 1969). This is assisted by the forking of the tongue, the tips of which, when inside the mouth, are close to the openings of the ducts of Jacobson's organs.

The relationship between the sizes of snakes and those of their prey has received considerable attention in recent years. Larger snakes tend to eat larger prey items and often add larger prey species to their diets. The range and variances in prey size also increase with the size of the snakes that feed on them. At the same time many species of snakes drop small prey items from their diet. This trend is particularly characteristic of fish-eating snakes. These matters have been discussed in relation to foraging theory by Arnold (1993). As already mentioned in Chapter 3, many nocturnal reptiles detect their prey by means of the chemo-sensitive organs of Jacobson, while the sensitive pits of Crotalinae respond to the warmth of its body.

The adaptations of extant snakes represent the current state of a process that had already begun in the Cretaceous era. Doubtless, this process of slenderization did not proceed under constant conditions. Not only have there been changes in the physical environment during the period, but both snake predators and their prey must have evolved as the success of snakes increased. Some snake taxa have concentrated on larger prey, others on numerous smaller food items; sometimes these trends have reversed. Slenderization has required solutions to four major problems, viz. how to maintain the patency of the trunk, how to maintain salt and water balance, how to control body temperature, and how to acquire adequate amounts of food. Some of these problems will be addressed in subsequent chapters.

Support and maintenance of the patency of the coelomic cavity is achieved in caecilians by hydrostatic pressure devices involving muscular partitions. Amphisbaenians, elongated lizards and snakes have articulated ribs that extend ventrally and stiffen the trunk, thus preventing excess pressure from being exerted on the coelomic contents. Possibly the most critical difference between snakes and other elongated squamates lies in the fact that the latter include relatively few food specialists, while the key to the success of snakes lies in specializations for capturing and devouring specific prey animals. Some of these specializations even preclude and restrict access to all but one type of food. For instance, snakes of the African genus *Dasypeltis*, which specialize on a diet of birds' eggs, exhibit slenderization of the skull and reduction in their dentition. These adaptations appear to be necessary for the ingestion of shelled eggs, but limit the capture of active prey (Gans 1983). Eggs are cracked in the throat by special bony spines that arise from the vertebrae and project through the lining of the gullet. Only the egg contents are swallowed, all fragments of the shell being regurgitated.

In the absence of fossil evidence, the evolution of the feeding movements of snakes is unclear, but Gans (1961) has postulated that liberation of the mandibular symphysis probably took place in the first instance. Freeing the tips of the mandibles allowed more complex movements of the jaw to be made, and unilateral feeding evolved from inertial feeding. Steps that make this possible are found in the nature of the muscle systems acting on the back of the jaws, and the connection of the tooth-bearing element in a simple lever pattern that initially kept problems of control to a minimum. Important secondary modifications resulting from these changes are the bony enclosure of the brain, the protrusible glottis (see below), constriction of the prey and the use of venom. The shift in feeding mechanics outlined above was probably only related indirectly to the loss of limbs. More important by far was the elongation and slenderization of the body.

The change to unilateral feeding mentioned above was accompanied by changes in the bony structure of the brain and the suspension of the quadrate bones from extended supratemporals. Dislocation of the jaws in this manner enables very large food items to be swallowed. In viperid or solenoglyphous snakes, erection of the poison fangs is brought about by two muscles which pull the pterygoid bones of the roof of the mouth forwards and upwards. These are connected via the ectopterygoids to the maxillae which bear the fangs and, in turn, are swivel-linked to be prefrontals. When the latter are pushed forwards by means of rotation, the fangs are erected. Both pterygoids and maxillae are capable of independent movement in relation to the quadrates, so the fangs can be erected individually when the mouth is opened (Fig. 59). (The term 'solenoglyph' is derived from the Greek words meaning 'pipe' and 'incision', and refers to snakes whose fangs are completely canalized and the suture scarcely visible. They have evolved the most efficient type of venom apparatus found among serpents.)

Snakes swallow their food whole and head first, after first immobilizing it by crushing or poisoning. The process is usually slow and may take hours. The jaws are so arranged that they can gape widely: the two halves of the lower

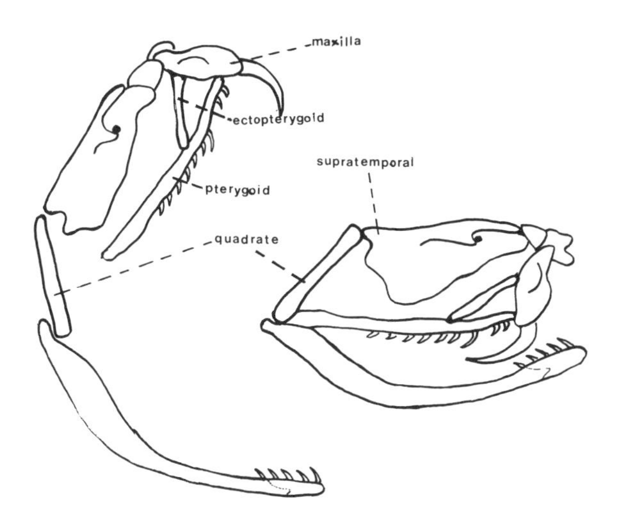


**Fig. 59.** Skull of viperid snake showing how suspension of the quadrate from the supratemporals enables large prey to be swallowed whole

jaw are connected by an extensible elastic ligament. All the teeth are recurved, and first one half of the lower jaw then the other is pushed forwards almost as though the snake were creeping over its prey. When the snake has passed its mouth over the victim, its neck may be enormously extended, every scale being separated. The cranial mechanics of snakes have been described in detail by Bellairs (1969) and are explained more simply by Bellairs and Carrington (1966). They have also been discussed by Engelmann and Obst (1984), Porter (1972) and many others.

The intimate association between the entrance to the respiratory tract and the entrance to the alimentary canal results potentially in respiratory difficulties when the mouth is used for prolonged processing of food. In typical reptiles, the choanae or internal nostrils – paired openings in the roof of the mouth which are connected to the exterior – are situated in a forward position so that the respiratory air stream passes through the buccal cavity on its way to the larynx. This makes respiration almost impossible when the mouth is filled with food, and various secondary adaptations have evolved that surmount the problem. In snakes, the larynx and trachea can be protruded from the mouth, like a snorkel, during feeding. A second solution, sealing off the respiratory air stream from the mouth cavity by prolonging the palate backwards, has been evolved independently in turtles, crocodilians, scincid lizards, and the therapsid ancestors of the mammals in which the development of a secondary palate paralleled the loss of kinetism and the differentiation of the dentition.

## 7.4 Venoms

Neither fishes nor aquatic amphibians appear to lubricate or process the food items in their mouths, and even the oesophagus shows few glandular regions. The maxillary gland of Anura, however, contains both amylase and protease enzymes. It has been suggested that these secretions form part of a functional complex. The glands discharge their secretions into ciliated grooves edged with taste papillae. In many species, these grooves curve around the patches of vomerine teeth, and the system may allow the animals to taste food objects after they have entered the mouth. Detection of prey by frogs appears to be primarily the visual perception of moving objects, with possibly occasional olfactory components in the process. Learning, and cues picked up by the tongue, may also be involved. Subtle decisions as to palatability are probably determined by assessment of materials dissolved from portions of the prey that have been abraded by the vomerine teeth. Enzymes have also been shown to occur in *Triturus* and *Salamandra* spp., but it is uncertain whether fossil terrestrial amphibians had similar enzyme-secretory systems.

In contrast, the buccal surface of turtles is keratinized and covered by stratified squamous cells, while deeper areas show mucous glands and organized glandular tissue. The buccal system of crocodilians shows only mucous-secreting glands and the function, likewise, is primarily one of lubricating the food. The mucous glands of lizards and snakes have become enlarged and organized . They produce a number of secretory products that lubricate and, in many cases, immobilize the prey.

Many reptiles are venomous to a degree: approximately one third of all species produce some buccal toxins, that is, substances that harm potential prey. Modified salivary glands that perform this function have been found in all the families of higher snakes, as well as in Lanthanotidae, Varanidae and Helodermatidae among the lizards. The widespread appearance of variably toxic venoms that are harmless to man in snakes and lizards indicates that venoms probably evolved in support of the capture and digestion of prey rather than as a deterrent to large predators (Gans 1978).

In the case of colubrid snakes, the primary function of the mucous glands is not for killing but for incapacitating the prey so that it can be ingested without difficulty (Gans 1978). This is especially important when the prey is so large that the jaws need to be 'disarticulated' and might be fractured if the prey were to struggle violently (Gans 1961). The true venom glands of elapid, hydrophiid and viperid snakes are located on the sides of the head and neck. Their ducts lie along the labial side of the maxilla and enter a complex pocket that surrounds the bases of modified maxillary teeth that are either deeply grooved or have a hollow tube within them down which the venom flows (Fig. 60). The oral glands of reptiles have been reviewed in great detail by Kochva (1978) and their evolution summarized by Underwood (1997).

Some hydrolytic enzymes are common to the pancreas, the mammalian salivary glands and the venom glands of snakes. The suggestion has therefore been made that, in their evolutionary development, venom glands first produced enzymes that were already being secreted by the pancreas and against which inhibitors were already present in the blood. These inhibitors facilitated the evolution of enzyme-based toxins by neutralizing any damaging substances that might have escaped through the venom glands (Kochva et al. 1983).

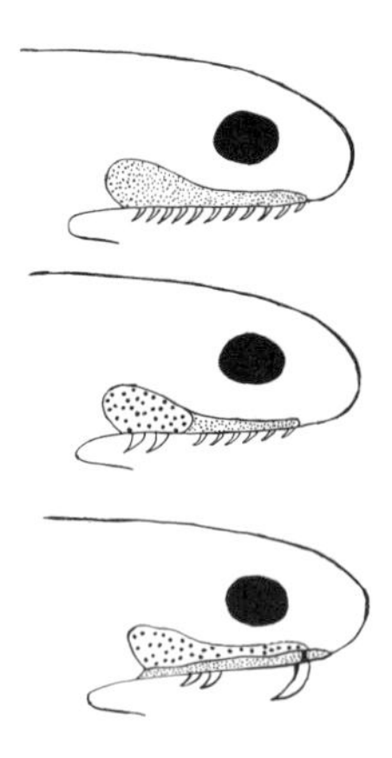

**Fig. 60.** Salivary and venom glands of snakes. *Above* Harmless colubrid showing salivary gland; *centre* back-fanged colubrid with rear part of salivary gland modified as a venom gland; *below* front-fanged snake with venom gland separated from salivary gland with a duct of its own discharging at the base of the poison-fang

The composition of snake venom varies greatly; indeed, only the venoms of closely related species are similar. In principle, the effects produced fall into two broad categories: one neurotoxic and causing damage to the nervous system, the other cytotoxic and heamorrhagic; that is, it breaks down tissue and destroys the erythrocytes or red blood cells. Neurotoxic venoms predominate in Elapidae - cobras, kraits, mambas and coral snakes - and in Hydrophiidae or sea snakes. Cytotoxic and haemorrhagic venoms, on the other hand, are characteristic of Viperidae, both Old World Viperinae and New World Crotalinae. The latter include rattlesnakes, copperheads, the water moccasin and the fer-de-lance. Many exceptions to this generalization are found, however. For example, the venom of the taipan (*Oxyuranus scutellatus*) (Elapidae) contains a considerable proportion of cytotoxic elements, while the tropical Cascabel rattlesnake (*Crotalus durissus terrificus*) and the Balkan adder (*Vipera berus bosniensis*) produce venoms that are mainly neurotoxic. In contrast to the last named, the common adder (*V. berus berus*) has a strongly cytotoxic venom (Engelmann and Obst 1984). The chemistry and immunology of reptilian venoms have been reviewed by Elliott (1978), the ecology and evolution of venomous snakes by various authors in Thorpe et al. (1997).

Even very large snakes seldom secrete more than about 5 ml venom, having a dry weight of 1.0–1.5 g. An eastern diamondback rattlesnake (*Crotalus adamanteus*) is credited with producing 6.25 ml venom, and even more has been taken from the largest specimens of fer-de-lance (*Bothrops atrox*) collected in Costa Rica; but these samples only yielded about 1 g dry venom. Minton and Minton (1971) tabulated the yields and toxicity of the venoms of numerous reptiles belonging to different families, and listed the most important enzymes found in them. They also outlined their biological activity and toxicity. (For a more up-to-date account of contemporary research on venomous snakes and their venoms see Thorpe et al., 1997.)

Within a single species of snake, the composition of the venom can show considerable geographical variation. Consequently, the bites of the same species in different parts of the world may produce different symptoms and require different treatments. The underlying causes of this phenomenon have never been explained, but recent work shows that the variation in the venom of the Malayan pit viper (*Calloselasma rhodostoma*) is closely associated with its diet. Adult pit vipers eat amphibians, reptiles, birds and mammals, but the proportion of each type of prey varies in different regions and the venom is appropriate for subduing and digesting the local diet (Daltry et al. 1996).

Bites by vipers are followed by severe pain and swelling in the region of the injury. Blood-stained serum may ooze from the fang marks and enter the subcutaneous tissues causing discoloration of the skin. Clotting of the blood is inhibited so that there may be haemorrhage of the lungs or intestine, and the patient coughs up blood or passes it through the rectum. Small purple spots often appear beneath the skin where the blood has leaked from damaged vessels. Later, areas of tissue become gangrenous and are sloughed away. When death occurs, it is usually due to failure of the heart or respiration. Although cobra venom not infrequently also causes severe local damage, there is little pain at the site of the injury.

The venoms of most snakes contain a variety of both neurotoxins, myotoxins and blood poisons, but the former tend to predominate in Elapidae, the latter in Viperidae, as we have seen. Moreover, neurotoxins may also act on the blood system, while blood poisons can have side effects on the nervous system, so that the effects of venom are complicated. Nevertheless, muscular weakness, as well as depression of the breathing and of the heart, are more characteristic of the effects of elapid venom, and death usually follows much sooner than in the case of viperid poisoning.

## 7.5 Metabolism

Amphibians and reptiles have the same general mechanisms of metabolism that are found throughout the animal kingdom. These are not merely those employed for capturing prey and utilizing energy, but also synthesizing, degrading and converting food and intermediate metabolites. The requirements of amphibians for specific metabolites, vitamins, cofactors, inorganic salts and metal ion activators are also much the same as those found in other vertebrate classes. The metabolism of Amphibia has been accorded detailed reviews by Brown (1964) and other authors in Moore (1964) as well as by Larsen (1992), the metabolism of reptiles by Bellairs (1969) and by Bennett and Dawson (1976).

Although reptiles show a wide diversity in their food, there is little variation in their resting metabolic rates both within and among extant taxa. In these, the rate of metabolism depends upon size and body temperature. The only exceptions appear to be xantusiid lizards and the tuatara (*Sphenodon punctatus*) (Rhynchocephalia), both of which have rates that are lower than would be predicted from their size. Moreover, with the exception of *Dipsosaurus dorsalis* (Iguanidae), all species so far studied obtain anaerobic energy at rates that are independent of body temperature. Activity is primarily supported by anaerobic means, supplemented by aerobic systems (Bennett and Dawson 1976).

The Burmese python (*Python molurus bivittatus*), like other Boidae, takes large meals, weighing up to a quarter of its own body weight, at infrequent and lengthy intervals. During the first 3 days after ingestion, its rate of metabolism in one experiment increased markedly: oxygen consumption by 17 times and intestinal nutrient uptake capacity by 6–26 times the fasting levels. The mass of the small intestine also doubled. The extra energy required for digestion was 32% of the total energy value of the meal, however, and much of it was expended before any energy had been absorbed from the prey (Secor and Diamond 1995).

Amphibians and reptiles clearly display a wider diversity in their diets than is generally supposed, and digestion of the different types of food ingested requires specialized adaptations. Insectivorous lizards and turtles secrete chitinase and chitobiase enzymes which aid in the digestion of the exoskeletons of insects and arachnids. Some herbivorous forms harbour populations of microorganisms that are able to digest cellulose, within an expanded portion of the alimentary canal (Skoczylax 1978). Heatwole and Taylor (1987) have discussed the feeding ecology of reptiles. They concluded that the relatively small number of herbivorous reptiles, especially Squamata, reflects the inability of these animals to chew or digest plant food effectively. Pough (1973) expanded this idea to suggest that larger lizards cannot satisfy their nutritional requirements on a diet of insects and other small prey, and therefore rely on vegetation. Among the Agamidae, Gerrhosauridae, Scincidae and Iguanidae, species with body weights exceeding 300 g are almost entirely herbivorous, whilst those weighing less than 50–100 g are carnivores. Moreover, juveniles of species that are herbivorous as adults tend to be carnivorous until they reach a weight of 50–100 g. The ecological consequences of the foraging modes of reptiles have been reviewed by Huey and Pianka (1981) and by Pianka (1986).

# 8 Reproductive Diversity and Life Histories

Most amphibians lay their eggs in water even though the adults may be completely terrestrial, while the vast majority of reptiles lay their eggs on land. When freshly laid, the eggs or ova are single cells, and development begins when they have been fertilized. Spermatozoa, formed in the testes of the males, are also single cells but greatly elongated and modified for locomotion. Ova and sperms are collectively known as gametes. Whereas the nuclei of somatic or body cells normally contain paired chromosomes, those of gametes are haploid and contain only half the number of chromosomes found in the somatic cells. Thus, if an animal species has $n$ chromosomes in its reproductive cells, the non-reproductive somatic cells will possess $2n$ chromosomes. This number is achieved at fertilization when the $n$ chromosomes of the egg fuse with the $n$ chromosomes of the sperm to form a diploid fertilized egg or zygote containig $2n$ chromosomes. This number is halved during the formation of gametes by meiotic cell division. Normal cell division is known as mitosis and, in this, the diploid state is retained, whereas in meiosis or reduction division each gamete receives only one member of each chromosome pair. Some species of amphibians and a few of reptiles are triploid having $3n$ chromosomes; even smaller numbers are polyploid and have more than $3n$ chromosomes.

The reproduction of higher animals usually involves a number of distinct elements, some of which may be curtailed or even dispensed with. These are as follows: (1) establishment of a territory, usually by the male; (2) courtship, in which both partners take part although the male frequently initiates the process while the female retains the right of refusal; (3) fertilization. This may take place externally after the eggs have been laid in water or internally either by copulation or by the female taking up a sperm packet (spermatophore) deposited by the male; (4) oviposition; and (5) protection of the eggs and young. The latter occasionally reaches a high degree of parental care. These elements of reproductive behaviour will be discussed in the following pages insofar as they are applicable to amphibians and reptiles.

## 8.1 Territorial Behaviour and Fighting

At one time it was not realized that many amphibians, like reptiles, may be territorial, but there is now a considerable amount of evidence for this. Territory is usually defined as an area within the home range of an individual that is defended against intruders. Indeed, it is only by watching the behaviour of a territorial animal in response to its competitors that the boundaries of its territory can be ascertained; and territory is best understood in the context of competition for limited resources. According to David Lack, writing in 1939 with special reference to birds, intraspecific fighting is the criterion upon which territoriality is based. Lack defined a territory as 'an isolated area defended by one individual of a species, or by a breeding pair, against intruders of the same species, and in which the owner of the territory makes itself conspicuous'. This definition can be applied equally well to fighting among both amphibians and reptiles. Within the home range of any adult animal, one or more areas in which mating, nesting, feeding or sheltering and so on take place may be defended and therefore qualify as territories. Males of many species are especially antagonistic to conspecific males, particularly in the breeding season, but both females and sometimes juveniles may, on occasion, also engage in territorial behaviour.

In most amphibian species, there is a marked disparity in size between the sexes, and males are usually smaller than females. This is because the females need to build up sufficient reserves of energy to produce quantities of relatively large eggs. Females apparently do not grow faster than males, but take longer to reach sexual maturity. In small and very small species, however, there is little or no difference in size between the sexes. This may be because, in such cases, the advantage gained from being able to make use of small, abundant food items outweighs the disadvantage of losing the capability, possessed by larger amphibians, of partitioning food resources between the sexes (Clarke 1996). Among those anurans that engage in physical combat, males are, again, often equal in size to, or sometimes even larger than, females.

Defence of territory by salamanders usually begins with the adoption of threatening postures. This is followed by an attack that involves lungeing, snapping, and sometimes even biting the opponent sufficiently violently as to cause it physical injury which occasionally includes loss of the tail. In laboratory studies, Thurow (1976) has recorded territorial defence, hierarchical social dominance and competition for feeding space between species of *Plethodon*, both large and small, in eastern North America. Adults of both sexes as well as juveniles may engage in such aggressive activities. The spacing of individual salamanders in the wild, as well as the stereotypical nature of much of their behaviour, suggest that aggression also occurs under natural conditions. The aggressive display of the red-backed salamander (*P. cinereus*) involves raising the trunk off the substratum and looking towards the opponent. A biting lunge often directed towards the tail or the nasolabial grooves may follow. A submissive individual lies flat on the substratum and looks away (Jaeger and Schwarz 1991). Other examples are cited by Stebbins and Cohen (1995).

Among anurans, spatial separation of males in breeding choruses is frequently observed, and vocal advertisement undoubtedly plays a part in such spacing (see below). Many terrestrial dendrobatid frogs defend territories that include calling sites, feeding sites, shelter and oviposition sites. In addition to vocalization, chest to chest wrestling and shoving matches sometimes occur, as well as jumping upon the back of an opponent, pushing it under water or off its perch, biting it, deflating its vocal sacs, and so on. Dramatic fights occur at their mud-pan nest sites between male blacksmith tree frogs (*Hyla faber*) in Brazil. These inclue pushing the sharp, curved rudiment of the thumb or pollex into the body of the opponent.

Physical defence of calling stations by resident males has also been described among Central American dendrobatid frogs. In *Dendrobates galindoi*, aggressive behaviour is signalled by elevation of the body and the emission of distinctive call notes. This may be followed by the opponents grappling one another with their fore legs. The poison frog *Prostherapis panamaensis* behaves in a similar manner at first, but physical attack takes the form of a charge, concluding with the resident male butting the intruder with its head. Other examples are cited by Salthe and Mecham (1974).

Temporal patterns of anuran reproduction fall into two broad categories: prolonged breeding and explosive breeding, as emphasized by Wells (1977) in a useful review. The distribution of females, both spatially and over time, determines the type of competition that takes place between males. The males of explosive breeders form dense aggregations and engage in competitive scrambles in which they attempt amplexus or pairing with any individual and struggle among themselves for the possession of the females. In contrast, the males of prolonged breeders usually call from stationary positions, thereby attracting females and often maintaining spacing among themselves. They defend territories, oviposition sites and courtship areas against conspecific males as already mentioned and, not surprisingly, the quality of their territores may enhance their attractiveness to females and thus enable them to obtain several mates in one season.

Sexual size dimorphism is found among reptiles as well as in amphibians. In most territorial species of Chelonia where males engage in combat with one another they are larger than the females. The same is true of most aquatic and bottom-living aquatic turtles and terrapins. Here combat between males is less common, but males forcibly inseminate the females. Both combat and forcible insemination are rare among truly aquatic species, and in these the males are usually smaller than the females. Whilst larger males may be more successful in winning fights, smaller males enjoy greater mobility and therefore are better able to locate females. Moreover, selection for fecundity may result in the increased size of females, which tends to counteract sexual dimorphism. In general, however, the extent by which males are larger than females increases with the mean body size of the species concerned (Berry and Shine 1980). Indeed, female chelonians are more often larger than males than vice versa. The greatest discrepancy is found in map and diamond-back terrapins (*Graptemys* and *Malaclemys* spp.).

In contrast to Chelonia, the largest crocodilians are usually males, although females are heavier than males of comparable length. Again, in lizards, the male is usually the larger sex. This is probably a general rule among territorial species, such as *Agama agama*. It is also true of the tuatara (*Sphenodon punctatus*) (Fig. 8). On the other hand, female snakes usually grow larger than their mates. The proportions and shape of certain parts of the reptile body may also vary according to sex: for instance, in many species of lizards and snakes the males have longer tails than the females (for details see Bellairs 1969).

Many extinct reptiles, particularly dinosaurs, had massive armour, horns, spines and other bizarre structures whose functions, until recently, have not been understood. They not only may have afforded protection from predatory enemies but also, almost certainly, were associated with intraspecific combat and ritual displays towards rivals of the same sex and with advertisement and courtship displays towards potential mates (Cloudsley-Thompson 1994). Primitive hadrosaurs were equipped with a small nasal horn which was used as a weapon in combats with members of their own species, and the dome-headed *Pachycephalosaurus grangeri* had a skull with a helmet 25 cm thick in front. Like *Stegoceras validus* (Ornithopoda) (Fig. 61) it probably fought duels, charging and butting as wild sheep do today. It seems probable that structures which originally evolve as weapons for intraspecific combat may later acquire a dual function and become defensive organs or ornaments for epigamic (sexual) display.

Territorial behaviour among extant reptiles is a well-known phenomenon. It has been recorded in chelonians, alligators and crocodiles (Lang 1989) as well as in squamates. The latter are known to defend their nesting sites and their basking places both during and outside the breeding season. The suggestion has been made that the unprovoked attacks by crocodiles upon small boats, which sometimes occur, may have been made in defence of territory.

One of the most thorough investigations of territorial behaviour among lizards is that of Harris (1964) on the rainbow lizard (*Agama agama*) in Nigeria. Each brightly coloured adult male or cock lizard shares his territory with one or more females and sometimes with several young. If another male in breeding colours appears, it is immediately challenged by the resident cock which attempts to drive it away. Vernon Harris described a range of gestures which are used by these gregarious lizards in appropriate circumstances. When confronted by a rival, a cock distends his gular fold and raises his head and fore-quarters several times. If the intruder holds its ground, the resident lizard raises its entire body up and down on all four legs. This threat is accompanied by fairly rapid colour change, the head changing

**Fig. 61.** Reconstruction of *Stegoceras validus* in sexual combat. (Cloudsley-Thompson 1994; after Halstead and Halstead 1981)

from orange to dark brown while the body becomes a pale bluish-grey. If this fails to daunt the adversary, a real fight develops. The two opponents stand side by side facing in opposite directions and lash at one another with their tails until one is knocked over and runs away (Fig. 62). Another gesture which consists of nodding the head is made by rainbow lizards to one another irrespective of sex or age. This may have some significance in group orientation.

Yeboah (1982) found that in two contrasting but adjacent areas at Cape Coast, Ghana, large territories held by large males associated with small numbers of young and subordinate animals were to be seen where food was plentiful and shelter scarce. In contrast, small territories occupied by small males and large numbers of young animals occurred where food was less plentiful but refuges were abundant. The difference could not be explained solely in terms of the weight of the males as those of similar size in the two areas held territories very different in size. The availability of food appeared to have little influence on density, but the abundance of refuges might have. It seems probable that a series of complex relationships exists between the quality of the habitat in terms of food, refuges, and competitive interactions between individuals within a group.

Territorial behaviour is well developed among *Anolis* spp. (Iguanidae) of North America. In these lizards, rapid extension of the throat-fan, which flashes bright pink as its skin is stretched, is a conspicuous feature of threat display. Male flying lizards (*Draco* spp.) (Fig. 16e) use their brightly coloured 'wings' as well as throat-fans in a similar manner as they run among the branches or glide through the air. Male marine iguanas (*Amblyrhynchus cristatus*) of the Galapagos Islands defend small areas of beach and indulge in ceremonial combats during which each individual attempts to overthrow its

**Fig. 62.** Rainbow lizards (*Agama agama*) fighting. (After Harris 1964)

rival by headbutting, but, as in the fights of *Agama agama*, the teeth are not used in aggression.

Elaborate territorial behaviour is found in chameleons, geckos and is most spectacular in varanids. Male *Varanus* spp. rear up and grapple each other with their front legs (Fig. 63) until one of the combatants is pushed over and allows himself to be chased away. Again, the teeth are not used, but monitor lizards may scrath each other's backs so severely with their claws as to cause bleeding. Male lacertids also threaten and fight with one another at the beginning of the breeding season.

Injuries such as broken tails, lost toes and various types of scars may result from intraspecific fighting. Many species of lizards have elaborate bebavioural signals that indicate the intent to defend their territories. Examples include distinction of the brightly coloured dewlap of *Anolis* spp. (Iguanidae), head bobbing in Agamidae, and so on. Done and Heatwole (1977) noted various degrees of aggression among Australian skinks, of which one of the most aggressive is *Sphenomorphus kosciuskoi*. Its displays consist of flattening and arching the neck and circling the opponent, and fights are vicious. *Ctenotus robustus*, which shows no aggression, lacks the social organization of *S. kosciuskoi* and *S. quoyi*.

Among lizards, herbivores, in general, are less aggressive than insectivores. The males either defend no territory at all or, if they do, it is small and subordinates are allowed within it. Females usually confine their aggression to the defence of nesting sites. Herbivorous males have a wider variety of mating systems than insectivorous males. These range from leks or small, defended territories (*Agama cristata*: Agamidae) to exploded leks (*Iguana iguana*:

**Fig. 63.** Male monitor lizards (*Varanus bengalensis*) in ritualistic combat. (After Bellairs 1969)

Iguanidae), and from large territories with subordinates (*Sauromalus obesus*: Iguanidae) to no territory at all *(S. varius*). The social behaviour of herbivores in the non-breeding season is more variable than that of insectivores. Herbivores may be highly clumped in dense aggregations (*A. cristata*) or thinly dispersed over wide areas (*Conolophus suberistitus*: Iguanidae) while clumps of lizards may move through the habitat (*I. iguana*) or have stable distribution (*Ctenosaura pectinata*: Iguanidae and *S. obesus*). It is reasonable to assume in lizards, as in other animals, that a comparatively predictable food source (arthropods) is more conducive to the evolution of territoriality than food resources, such as flowers, fruit and new leaves, that fluctuate in distribution and abundance. The general passivity of herbivores compared with insectivores may well be directly attributable to difference in diets (Stamps 1983). While many species of lizards are territorial, others are hierarchial and some have harems. For all those territorial species studied, crowding results in increased social interaction, increased aggression, and a switch to hierarchial behaviour. The evolution of social behaviour in reptiles has been reviewed by Brattstrom (1974). The widespread occurrence of territoriality in lizards is concerned with reproduction and mating, food and foraging, basking and thermoregulation. This subject has been discussed by Stamps (1983), Heatwole and Taylor (1987) and Martins (1994), among others.

Although it is not known with certainty whether or not snakes are territorial, ritualistic combats or snake dances have been recorded between male rattlesnakes, vipers, mambas, copperheads and various colubrids. These all entwine their bodies and, in some species, rear up and push against each other (Fig. 64) until one of the two gives up and glides away. These combats do not usually cause any serious injury since the snakes make no attempt to bite each other. So far, snake dances have not been satisfactorily explained. Only males are involved and they usually fight in the mating season. However, they may do so when no female snake is in the vicinity and, moreoever, as already stated, there is no evidence that male snakes lay claim to particular territories or are prepared to defend them (Bellairs 1969).

## 8.2 Vocalization

Sound production is somewhat limited in salamanders and caecilians, as it is in reptiles, but is highly developed among anurans. Vocalization is a method of advertising the presence of one individual to others of the same

**Fig. 64.** Combat between male rattlesnakes (*Crotalus atrox*). (After Bellairs 1969)

species, and it may function both for the attraction of mates and in the establishment of dominance hierarchies among males. Duellman and Trueb (1986) give a classification of the known kinds of anuran vocalizations and their functions. These include: (1) advertisement calls emitted by males, both in courtship, territorially in response to the calls of the other males above a critical threshold of intensity, and as encounter calls evoked at close range during aggressive or agonistic encounters between conspecific males; (2) reciprocation calls given by some receptive females in response to the advertisement calls of males; (3) release calls; these are associated with vibrations of the male's body or by unreceptive females in response to amplexus; (4) distress calls produced by both sexes, usually with the mouth open, in response to disturbance. Other authorities also recognize rain calls and warning calls.

Approximately 2600 species and subspecies of frogs and toads survive in the world today. Most of them are vocal and possess distinct repertoires of sounds. The total number of calls in the vocal repertoire varies from species to species. Some anuran species produce only one or two, others several different calls. The extent of the repertoire reflects the selection pressure for promoting reproductive isolation within the species, the ecological environment that it inhabits and the development of its social behaviour (Capranica 1976).

Male anurans often compete for mates using alternative tactics whose relative success may be influenced by four factors: subsequent behaviour, physiological state, the frequency of expression of alternative tactics and the density of competing males. These alternative tactics include female minicry, forced copulation, cuckoldry and satellite behaviour (Arck 1983). Satellite males associate themselves with males that are actively advertising themselves to attract females. Then they attempt to intercept and mate with any female that approaches the advertising male. According to Lucas et al. (1996) who analysed chorus behaviour as a stochastic dynamic game, satellite behaviour is less common when the length of the breeding season is constrained by extrinsic factors such as seasonal weather patterns. Extrinsic constraints may also affect the stability of the system. This may only continue for a short time if predation rates are sufficiently high, or if satellites intercept a large fraction of incoming females. Thus, explosive breeding systems can result from either biotic or abiotic causes.

Compared with that of anurans, the vocalization of reptiles is rare and undistinguished. A notable exception is, however, provided by the ground-living geckos (*Ptenopus garrulus*) of the Namib and Kalahari deserts. These noisy little creatures emerge from their burrows in the sand just before sunset and begin a chorus of chirping that ends abruptly when darkness falls. It is extremely difficult to detect individual geckos, however, because, like cicadas, they become silent when approached and it is almost impossible to detect one call from another. The significance of this chirping is not clear, but it probably represents some form of territorial behaviour.

## 8.3 Courtship and Fertilization

As in many other animals, much of the social behaviour of amphibians centres around competition among males for access to mates. Because of the low cost of sperm production, compared with that of eggs, males are potentially able to mate with many different females. The location and stimulation of potential mates is primarily the task of male amphibians, but, in some species, females also play an active part. Courtship among amphibians has been studied extensively. In the majority of urodeles, such as the smooth newt (*Triturus vulgaris*), a complex sequence of interactions takes place between the male and the female (Halliday 1975). Such interactions are highly variable, but there are three main stages. The first, one of orientation, takes place when the two sexes meet one another. The second is mainly one of static display in which the male shows his flank, adorned with its crest, to the female and then moves up and down. Finally, the male fans a steady stream of water containing hedonic glandular secretions or 'pheromones' towards the female. Eventually he backs away and deposits sperm packets or spermatophores on the substrate. The female takes these into her vent where the spermatozoa are liberated and swim into the spermatheca. Here they are stored sometimes for up to 2 years until the time of oviposition or egg laying (Frazer 1983; Griffiths 1996). The spermatophores are deposited in water, on land or both, depending upon the species. The large American hellbender (*Cryptobranchus alleghaniensis*) is unusual among salamanders in that the female lays her eggs in a nest excavated by the male in the muddy bottom of a river. Several females may lay in succession in the same nest, the male emitting sperms after each. He then guards the eggs until they hatch a couple of weeks later.

The pattern of courtship among newts and salamanders varies with the species, but the secretions of the hedonic glands play an important role in all cases. These scent glands are found in different parts of the skin, but are especially numerous around the face and cloaca. During the display of the male, pheromones are wafted towards the female when the former lashes the water with his tail. In some newt species, the male acquires gaudy nuptial colours which, in the first instance, attract the attention of the female, but the important phase is when the male exposes the female to the secretions of his hedonic glands and deposits a spermatophore. This is placed on the bottom of the pond or stream and the female moves forward and takes it up whth her cloaca.

Reconstructing the phylogeny of behaviour patterns is highly speculative, partly because of the inevitable lack of fossil evidence, and partly because behaviour appears to be very plastic on an evolutionary time scale. In the case of salamanders, for instance, courtship is notably complex at two levels.

Many species have elaborate courtship with complex temporal structure and there is considerable diversity of courtship behaviour within the Urodela. This elaborteness is largely a response to the difficulties of indirect sperm transfer. Species that practise external fertilization (Cryptobranchidae and Hynobiidae) show relatively simple sexual behaviour, and the activities of the male are directed towards eggs rather than towards females. Salamanders have organized their sexual behaviour around spermatophore insemination over millions of years, however, and the behavioural radiation we see today reflects commitment to an intrinsically difficult mode of fertilization. Many archaic courtship activities probably represent important solutions to adaptive problems. For instance, the tail-straddling walk of the Plethodontidae and the right-angle turn of salamandrids are major innovations in the process of sperm transfer. Both lessen the chance that receptive females will fail to find the spermatophore. Diversification of courtship techniques is a striking feature in the behavioural ecology of salamanders and can be interpreted as a consequence of rapid co-evolution of the ability of males to persuade the females to mate, coupled with their own responsiveness (Arnold 1977).

Male anurans do not acquire nuptial colours in the breeding season as newts do, but they develop swellings on the fingers or forearms which are known as nuptial pads and assist in clasping the female who has been attracted by their vocal calls. The shape and size of these pads are normally characteristic of the species but may vary in different parts of its range. After

**Fig. 65.** Various species of anurans in amplexus

the female has been grasped in amplexus (Fig. 65) the eggs are extruded and sperm ejaculated over them. Different species of both Urodela and anura show a great diversity of reproductive patterns, for details of which the reader is referred to the books by Taylor and Guttman (1977), Duellman and Trueb (1986) and Stebbins and Cohen (1995) in particular.

Some aspects of caecilian reproductive biology are unique among amphibians, others among vertebrates generally. All known species have an intromittant organ, the posterior part of the cloaca (Fig. 66). This is everted by a combination of contractions of the body wall and of contractions of the musculature of the cloaca itself. Other vertebrates, including birds, evert part of the male cloaca during copulation, but not to the extent and complexity seen in caecalians. Internal fertilization in all species of the order is in maked contrast to the frogs, of which very few species appear to practise it, and to the urodele situation in which primitive species practise external fertilization. Only the unique North American tailed frog *Ascaphus truei* copulates by means of its so-called tail, an intromittent organ, while the few other genera, such as *Nectoprynoides*, that do have internal fertilization apparently transfer sperm by cloacal apposition. The intromittent organ of *A. truei* is a structure comprised of muscles and cartilages retained from the tadpole stage. It is inserted into the vent of the female during amplexus, when the male grasps the female with his fuselage. Among most salamander species, internal fertilization is achieved by the male depositing a spermatophore which is picked up from the substrate by the female using her vent (Wells 1977).

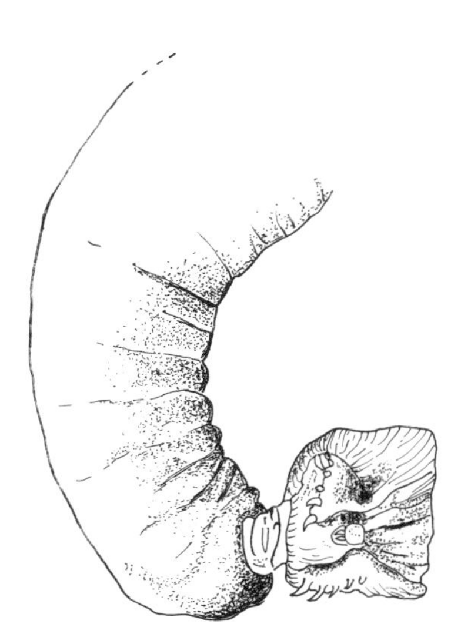

**Fig. 66.** Everted intromittant organ of a male caecilian. (After Noble 1931)

Whereas courtship is mainly chemical in Urodela and vocal among Anura, in reptiles it tends to be more dramatic. Marine and freshwater chelonians mate in the water and, in some species, there may be quite elaborate patterns of courtship. Male terrapins, such as *Pseudemys scripta* and *Chrysemys picta*, swim backwards in front of the females, tickling their chins with elongated front claws, while African side-necked turtles (*Pelomedusa subrufa*) spend long periods under water apparently kissing, and male terrestrial tortoises butt the carapaces of the females with their own as they butt those of rival males. The male of the large African species *Geochelone sulcata* makes loud gasping noises, which are audible for some distance, after mounting the carapace of the female and inserting his long, curved penis into her cloaca.

Bull alligators and crocodiles roar, which sounds like distant thunder, and this is believed to attract the females. Mating takes place under water. Much of the territorial behaviour of squamates, described above, probably also functions as courtship when directed towards females. Although snakes have fewer social gestures than lizards, the two sexes of some species nod towards one another. Male boas and pythons use their rudimentary hind limbs for scratching the flanks of the female, making an audible rasping sound. This persuades the latter to move herself into a position in which he can twist his tail under hers and insert his hemipenis. Before mating, male snakes often follow the females around, flicking out their tongues and running their heads and necks over the female's body. The king cobra (*Ophiophagus hannah*) nudges the extended hood of the female with his snout at the same time. In many Colubridae such as *Thamnophis* and *Storeria* spp., the body of the male is thrown into a series of waves passing from head to tail. In rattlesnakes, on the other hand, the male thrusts or jerks his body instead of undulating it. Tactile and chemical senses play a major part in the sex life of snakes whereas, among lizards, vision is more important (Noble 1937; Bellairs 1969).

Some 32 taxa of reptiles have been tentatively identified as parthenogenetic. That is, their eggs develop without being fertilized. Parthenogenesis among reptiles appears to originate in hybridization. Although this has not yet been completely proved, no exceptions have so far been found to the generalization. Parthenogenesis occurs in groups in which certain kinds of genetic differentiation have occurred so that the hybrids are automatically parthenogenetic (Darevsky et al. 1985). In some animals, parthenogenesis with only females in the population alternates with ordinary sexual reproduction, which allows for the recombination of genetic material and requires fertilization by a male. When the eggs of *Lacerta saxicola* are maturing in the ovaries of parthenogenetic females, they undergo normal reduction division. The next cell division is never completed, however. Instead, the nuclei - each containing the haploid number of chromosomes - fuse instead of passing into respective daughter cells. Consequently, a diploid set of chromosomes is formed. If a parthenogenetic female mates with a normal male, the sterile female hybrids which result are triploid (see discussion in Bellairs 1969: parthenogenesis has also been reviewed by Darevsky et al. 1985).

## 8.4 Oviposition

Newts (*Triton* and *Triturus* spp.) deposit their eggs singly, which is advantageous in standing water, while salamanders that breed in running water (the families Hynobiidae and Cryptobranchidae) attach their egg sacs to rocks and aquatic plants. Here they are aerated, protected and prevented from being washed away (Schmalhausen 1968). Among anurans with amplexus, the eggs are laid in batches and fertilized immediately as they are extruded. In some Pipidae, complex manoeuvres take place under water and there is an evolutionary trend from laying the eggs on plants in midwater to ovipositing upside down on the surface of the water, when the eggs remain floating.

Tropical frogs not infrequently lay their eggs completely out of water. Some lay them on forest leaves or on the branches of bushes or trees. Sometimes they are laid in places from which they can be washed into standing water after the next heavy fall of rain, or on leaves overhanging the water so that, on leaving their eggs, the larvae drop into it. In other species, the eggs are placed in damp situations on leaves and hatch as tadpoles or immature air-breathing frogs. Oviposition is accomplished in much the same way on land as in water, except that in species that lay their eggs on land, the male maintains close contact with the back of the female and their vents are continuously juxtaposed. The female in some species, with the male still in amplexus clutching her back, descends to a pool or stream beneath the vegetation and takes water into her bladder. She then returns to the vegetation, deposits a clutch of eggs, and releases the water over them as they are laid. Subsequent clutches are deposited by the same pair of frogs, but only after descending to the water again. Female *Phyllomedusa* spp. do not descend to water before laying their eggs. Some species, however, fold a leaf over them as they are being deposited on it. Parental care by amphibians has been reviewed by Ridley (1978) and more recently by Duellman and Trueb (1986) and Stebbins and Cohen (1995).

The ancestral mode of reproduction in reptiles is oviparity or the deposition of a fertilized amniotic egg, which is usually large-yolked or megalecithal. The young are precocial and, in most species, require little or no parental support after hatching. Ovoviviparity (see below) is not uncommon, however, among lizards and snakes, being found in about 15% of all extant species. Reptile ovoviviparity ranges from simple egg retention to the development of placental structures and reduction of the yoke. Its diversity indicates that it evolved independently in a number of taxonomic groups (Shine and Bull 1979). It is frequently seen among squamates, mainly in cool climates, but the selective advantage of uterine retention of eggs is somewhat obscure. It might perhaps enhance survival by protecting the developing young; but Shine (1995) has recently suggested that a more likely answer to the problem is that it enhances hatchling viability directly – because eggs incubated at maternal body temperatures are fitter than those that develop at normal nest temperature. The young are larger, can run faster, and their anticipated defences, for example, are better developed, as is their propensity to bask in sunlight. In addition, viviparity may be significant in allowing incubation of the eggs at high altitudes and latitudes where low temperatures prevail. Nevertheless, information on the temperatures required for embryogenesis among reptiles is still somewhat meagre (see Cloudsley-Thompson 1971).

The number of eggs laid is related not only to the absolute size of the reptile that lays them but also to the shape of its body. Griffith (1994) showed that in the scincid lizard genus *Brachymeles*, relative abdominal size is inversely related to its elongation, suggesting that relative clutch mass decreases with greater abdominal body length. Because few elongate lizards have near relatives that retain an elongated shape among larger species, it is seldom possible to assess this relationship, but *Brachymeles* is somewhat unusual in this respect. Griffith's study confirms that constraining factors include the retention of anterior body segments bearing parasternal ribs (which prevents extension of the clutch anteriorly within the body) and reduction of allometry of the abdominal segments. This provides extended series of uniformly sized vertebrae for limbless locomotion, but reduces the relative size of the abdomen, which affects relative egg size, number and packing in the available space.

Reproductive traits in the mammal-like reptiles probably included retention of the eggs, ovoviviparity (explained below) and maternal care – which included nourishment of the young by brood-patch secretions. (Oviparous, egg-laying forms also occurred, presaging the monotremes – echidnas and the duck-billed platypus – of today.) Viviparous therapsids would probably have possessed a placenta that functioned in fluid exchange and possibly in the supply of nutrients. Hypervascularity of the oviduct and embryonic membranes occurs in a variety of egg-retaining oviparous reptiles today, presumably in response to oxygen shortage or hypoxia and the prevailing hormonic conditions. The sequence probably began independently many times among the Therapsida and possibly among the pelycosaurs as well, since it did not require endothermy. According to Guillette and Hotton (1986), endothermy would have enhanced the advantages of placentation in all its evolutionary stages, each of which was functionally interrelated but independent of endothermy. Egg guarding and nest care probably gave rise to true incubation of the eggs or brooding of the nestlings around the end of the Permian. Maternal care of the nestlings would have followed, especially if pheromones from the brood patch prolonged the association between mother and young.

Most reptiles lay eggs, but some lizards and snakes give birth to live young by retaining the eggs within the oviducts until they are fully developed. This is known as ovoviviparity, a phenomenon already mentioned. Shine and Bull (1979) proposed that live-bearing evolves in cold environments (see Chap. 10) and that maternal care of the eggs facilitates its evolution. They analysed data from over 1000 species, revealing that live-bearing has evolved recently at least 38 times, while an additional 60 species show intermediate degrees of egg retention. Species that have recently evolved live-bearing show the intermediate stages and are indeed more often found in cold climates, or show greater maternal care of the egg than squamates in general.

As in the length of mammalian gestation, there is a correlation in reptiles between neonatal brain mass and incubation time. The squamates can be divided into at least three major groups on the basis of this as follows: Serpentes with the shortest incubation times, the Iguanidae, Agamidae and other groups dominated by small members in the middle, and teiid and varanid lizards showing the longest incubation times. The distribution of ovovivarity within the squamate reptiles is an implication of these findings. Most current hypotheses on the development of live-bearing suggests that its evolution was driven by the need to speed development in cold climates through maternal thermoregulation during pregnancy, protection of the developing young by large or venomous mothers, and so on (Shine and Bull 1979; Shine 1985, 1995). Such reproductive behaviour may, however, come at the cost of increased likelihood of predation as a consequence of reduced mobility (Shine 1985). If egg incubation times are longer in some groups than others, it should increase the cost of such behaviour, this cost being the increased likelihood of predation over a longer period, reduction in the number of clutches per reproductive individual and the costs of pregnancy. Consequently, the longer incubation times in larger or large-brained lizards probably preclude the development of ovoviviparity in these animals. That viviparity is generally absent among the larger lizards and those with the longest incubation times - the Teiidae and Varanidae - is often observed in Boidae and other snakes of similar or greater size, and is consistent with this hypothesis (Birchard and Marcellini 1996).

## 8.5 Competition and Cannibalism in Amphibia

After hatching, tadpoles frequently compete with one another in a variety of direct and indirect ways. The effects of crowding on the suppression of growth among anuran tadpoles have been the subject of laboratory studies for decades. Since the publication of Richards' (1962) pioneering work, unicellular organisms have been implicated in competitive interference between tadpoles of several different species, although tadpoles of the southern leopard frog (*Rana utricularia*) and those of the wood frog (*R. sylvatica*) in the United States, for instance, have little effect upon one another under experimental conditions. Growth inhibition of small natterjack (*Bufo calamita*) tadpoles by large larvae of *Rana temporaria* have been demonstrated in England by R. A. Griffiths and co-workers. These authors used replicated outdoor ponds which closely resembled many natural pools used by these species (Griffiths et al. 1991, 1993). They found that a unicellular organism in tadpole faeces was responsible for a high level of interspecific competition which decreased when the breeding periods of the two species were well separated. This organism, an unpigmented alga (*Protothica* sp.), was first identified by Beebee and colleagues (see Beebee 1995 for a resumé of the evidence of an interference effect in nature despite the inconsistent results of much recent research).

The complex life cycles of salamanders and anurans create a complicated web of intraspecific interactions that often include cannibalism. If this is defined as killing and eating live conspecifics, the almost universal consumption of dead animals and parts of their bodies such as tails, limbs and gills can be excluded from discussion. Cannibalism is widespread among amphibians. It has been reported in 7 of the 9 families of salamanders, 12 of the 21 families of anurans, and in 1 out of 5 families of caecilians (Crump 1992). A common form of cannibalism, practised by all developmental stages of amphibians, but mostly by larvae, is oophagy or egg-eating. Eggs provide a rich source of energy, calcium and phosphorus. One type of oophagy is opportunist and does not involve any specialization for eating eggs: the second involves eating unfertilized nutritive or trophic eggs provided by the mother of the tadpoles. Almost all species known to exhibit trophic feeding are arboreal and live in the water-filled axils of plants or tree holes. An exception is afforded by the African *Afrixalus fornasinii*, a terrestrial species whose unique leaf-gluing mode of egg deposition may perhaps serve as a defence against interspecific egg predation (Drewes and Altig 1996).

The most frequent type of cannibalism among amphibians is the eating of larvae by other larvae of the same species. The larvae of the salamanders *Ambystoma tigrinum*, *Salamandra salamandra* and *Hynobius nebulosus* are particularly cannibalistic, but Crump (1992) provides a table of numerous species of Caudata and Anura that have been recorded practising cannibalism in the field. Cannibalism can have significant implications for the ecology of amphibians, especially in overcrowded situations where food resources are limited, by increasing growth rates and decreasing the time of development to metamorphosis. Species with appropriate jaw morphology and digestive systems most frequently become cannibalistic when conditions are appropriate. In some populations of *Ambystoma tigrinum*, for instance, three larval and adult morphs, or forms, occur. A large morph inhabits permanent ponds and does not metamorphose. It is neotenous and becomes sexually mature whilst a larva. The second morph is smaller and inhabits ephemeral ponds where metamorphosis takes place. Some smaller morphs, however, develop proportionately larger heads, wide mouths, elongated teeth and prey on conspecific larvae. They grow rapidly and metamorphose early (Duellman and Trueb 1986). They benefit especially in temporary pools that dry up rapidly whereas the morphs with normal feeding habits are more successful when a longer period of development is available. The production of cannibal morphs is an insurance in seasons of low rainfall.

## 8.6 Parental Care

In addition to providing trophic eggs for cannibalistic larvae, nursing habits are very common among frogs, as Gadow (1901) emphasized long ago. In many cases the eggs are carried about during their development by the male or female parents. A classic example is afforded by the marsupial frogs

(*Gastrotheca* spp.) in which the developing eggs are carried around in pouches on the female's back (Frazer 1973; Fogden and Fogden 1988). In *G. marsupiata* and allied species, the skin of the back is studded with numerous calcareous plates so that the developing young are safely enclosed within a veritable coat of mail (Noble 1931).

The tadpoles of arrow-poison frogs (*Dendrobates* and *Phyllobates* spp.) develop in bromeliads, or in small holes in tree-trunks containing a few millimetres of water only. They are carried between these reservoirs of water on the backs of females and males. In such cases, the tree frogs remain safely in the higher vegetation of tropical forest, and need never descend to surface waters to breed, while the eggs and tadpoles are kept away from the many predators which might otherwise feast on them. Presumably the latter are less toxic than the adults, otherwise the argument would be paradoxical. The eggs and tadpoles of some Hylidae such as *Hemiphractus* spp. develop in a dorsal brood pouch of the female (Fig. 67).

Foam nests are produced by a number of frogs of the genera *Rhacophorus*, *Leptodactylus* and *Physalaemus* (= *Engystomops*), etc. They may be found in dry hollows on the ground, along the banks of streams, or even aboveground and attached to bushes or the lower branches of trees. The tadpoles remain in them for a while after hatching until the nest liquifies. Downie (1988) analysed the functions of foam nests of the common leptodactylid *Physalaemus pustulosus*. Unlike Gorzula (1977) and Dobkin and Gettinger (1985), he concluded that the foam has little thermal significance. It may, however, help to

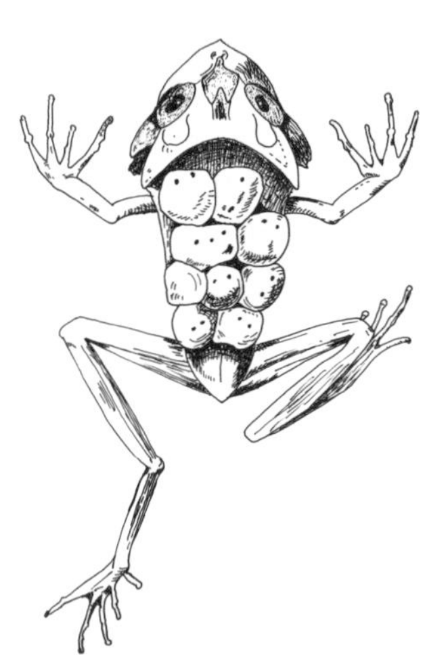

**Fig. 67.** Female *Hemiphractus bubalus* (Hylidae) carrying developing eggs on its back

protect the eggs and hatchlings from desiccation. It is also likely that it may have anti-predator properties.

This demonstrates the problem of why so many tropical and especially neotropical anurans reproduce on land. There are probably a number of adaptive advantages, operating simultaneously. Rivers and their permanent streams apart, lakes, ponds and standing water generally are rare in the rainforest. At the same time, puddles and temporary pools in the tropics speedily acquire a rich fauna of predatory beetles, water-bugs, dragonfly and predaceous dipterous larvae. It may well be that the avoidance of insect predators is a more important selective factor than the lack of standing water. Certainly, insects are often extremely resistant to toxins that have a remarkably harmful effect upon vertebrates. It seems probable, therefore, that the selective advantage to amphibians of reproducing on land rather than in water, in the tropics, may lie in the avoidance of predatory aquatic insects that are resistant to amphibian skin toxins, such as batrachotoxin, which have a paralysing action on, and are so effective as defences against, vertebrate enemies.

Parental care is complex in dendrobatid frogs, all of which are day-active and transport their tadpoles, whose venter adheres to the dorsum of the male or female by a sticky mucus. Males of the two species of *Rhinoderma* (Rhinodermatidae) transport tadpoles in their vocal sacs after picking them from the deteriorating jelly of the terrestrial egg clutch. In the case of the midwife toad (*R. darwinii*) they complete their development there and emerge as froglets.

Parental care has evolved repeatedly in anurans (Gross and Shine 1981). In some species it is performed by the male, in others by the female. The prevalence of male care is probably correlated with external fertilization resulting from male territoriality and this, in turn, is a consequence of female discrimination between oviposition sites. Internal fertilization, in contrast, preadapts females to selection for embryo retention, leading to live-bearing.

Males of the Jamaican frog *Eleutherodactylus cundalli* are territorial and call from exposed rock sites, but the eggs are laid in caves and other sheltered sites. The females attend the clutch, apparently protecting the eggs from fungus infection and predatory land crabs. When the eggs have hatched, the froglets climb onto their mother's back and are transported into the open. *Eleutherodactylus cundalli* is the first frog species reported to breed in caves, a habitat that has a stable environment, moderate temperatures and high humidity (Diesel et al. 1995). (For accounts of recent research on social behaviour in amphibians see the articles in Heatwole and Sullivan 1995.)

Like ovoviviparity, parental care by reptiles enhances survival of the eggs and young. The more primitive members of the Synapsida were probably obligatory egg-layers and guarded their eggs and nests, showing a range of behaviour comparable with that found among some extant reptiles (Guillette and Hotton 1986). (Herding in vegetarian dinosaurs has already been discussed in Chapter 7.) Then there is the fossil evidence for communal nesting among hadrosaurian dinosaurs which showed many similarities with that of *Iguana iguana*. Green iguanas use the same nest sites repeatedly. The mothers guard the nests, the eggs are laid at the beginning of the dry season so that drowning of the eggs is avoided, and hatching takes place with the onset of the rainy season when fresh, nutritious food becomes available.

The ceratopsian *Protoceratops andrewsi* was another ornithischian dinosaur known to have required parental care in order to survive (Fig. 68). In fact, there is a limit to the size which an egg can reach. If the shell is too thick and impervious, the developing embryo within it will suffocate; if it is too hard the embryo will not be able to hatch. This sets a geometric limit to the size of eggs, which is approached when they reach such a volume that the pressure of the internal fluid exceeds the strength of the shell that contains it. It is probably for this reason that the largest dinosaur eggs were oval and with

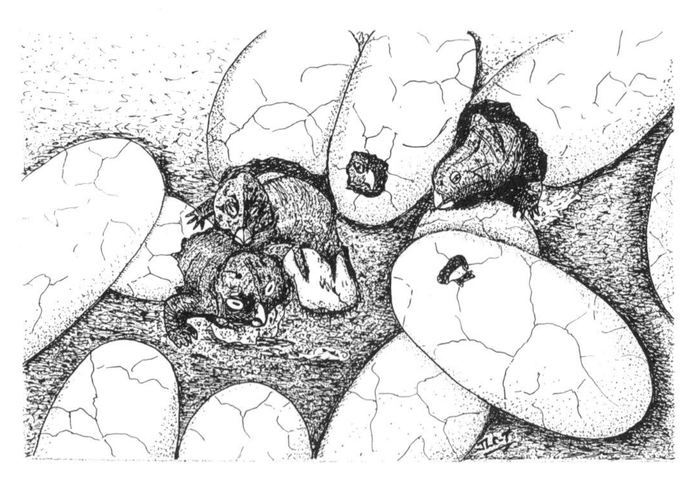

**Fig. 68.** Reconstruction of a clutch of dinosaur (*Protoceratops andrewsi*) eggs hatching, Late Cretaceous. (Cloudsley-Thompson 1994)

a somewhat flattened shape which spread their weight on the ground (see Chapter 2). With their relatively small size and naked skins, baby dinosaurs would almost certainly have needed parental care in order to survive. Some dinosaurs evidently aggregated not only for breeding purposes but throughout their lives, so they may well have been truly social (Fig. 69; see Cloudsley-Thompson 1978b, 1994).

The recent discovery of a fossil oviraptorid theropod (*Oviraptor*) lying on a nest that contained about 22 eggs suggests that some dinosaurs may have evolved avian-type nesting and incubation of the eggs. Of course, the animal in question could have been robbing the nest of another dinosaur at the time of its death. *Oviraptor* lacked teeth for biting prey, but two short teeth pointed downward from the roof of its mouth which might have been used to crush the shells of other dinosaurs' eggs in the same way that a sharp vertebral process in the neck helps egg-eating snakes (*Dasypeltis* spp.) to cut neatly through the shells of the birds' eggs that they swallow. It appears more likely, however, that the animal in question was incubating its own eggs, for the front limbs were directed posteriorly, with both arms wrapped around the nest. Another possibility is that the animal perished whilst in the act of ovipositing in the nest (Fig. 70). This seems to be obviated, however, by the absence of eggs inside the body cavity (Norrell et al. 1995).

Chelonians bury their eggs comparatively deeply in sand or soil and then leave them to their fate. Marine turtles only come ashore during the breeding season. Most oviparous squamates, in contrast, lay their eggs under superficial cover on the surface of the ground in places where conditions of temperature and humidity are suitable for development. Many crocodilians bury their eggs in sandy places, but the American alligator (*Alligator missippiensis*)

**Fig. 69.** Reconstruction of a herd of brontosaurs (*Apatosaurus ajax*) (length ca. 21 m), Jurassic. (Cloudsley Thompson 1994)

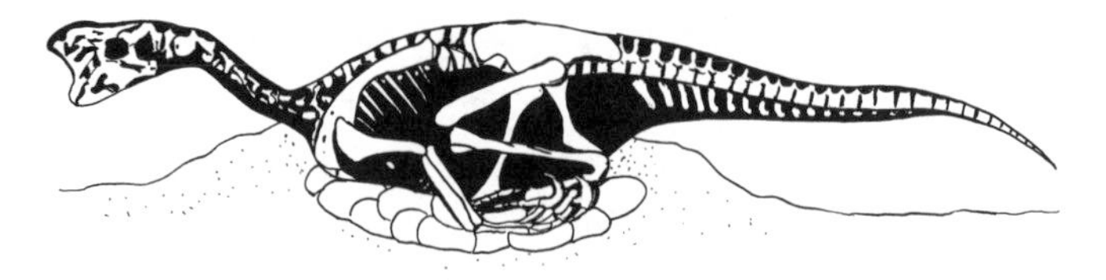

**Fig. 70.** Reconstruction of *Oviraptor* on its nest shortly before death. (After Norell et al. 1995)

builds a nest almost entirely of vegetable matter and the estuarine crocodile (*Crocodylus porosus*) constructs one of vegetation mixed with mud. The temperature at which the eggs develop determines the sex of the hatchlings, as we shall see in Chapter 9.

Care of the eggs is widespread among extant reptiles, but, except in crocodilians, protection of the young is exceptional. Both Aristotle and Pliny described how Nile crocodiles (*Crocodylus niloticus*) guard their nests. When the babies are ready to hatch, they make croaking sounds inside the eggs. Their mother hears, and digs them out of the ground. The young could not escape otherwise because the nest material becomes very hard during the period of incubation. Once the eggs have been uncovered, however, the babies are able to break through the shell with the aid of a flattened and horny egg tooth, on the top of the snout, which drops off soon after hatching (Fig. 71). The mother and father crocodile then lead the babies to the water, sometimes carrying them in their mouths (Fig. 72). Shine (1988) presents a table of crocodilian taxa reported to show parental care. Although the amount of attendance and defence of the nest varies interspecifically and even within species, the broad details of parental care appear to be similar in all crocodilian species and quite different from those described in any squamate.

With one or two possible exceptions, there is little evidence among lizards of the defence of eggs or repulsion of egg predators, but, in snakes, there can be a marked degree of parental care, especially among large and venomous species. In the king cobra (*Ophiophagus hannah*), for example, both sexes help to build a nest among bamboo stems and then guard it. Shine (1988) also tabulated the occurrence of parental care among squamate taxa as he did in the case of crocodilians. In discussing its evolution, he concluded that the proportion of species with parental care is lower among reptiles than among amphibians. This difference reflects the absence of male parental care in squamates, which, in turn, may be due to internal fertilization.

**Fig. 71.** Hatching Nile crocodile (*Crocodylus niloticus*) showing egg tooth. (Cloudsley-Thompson 1994)

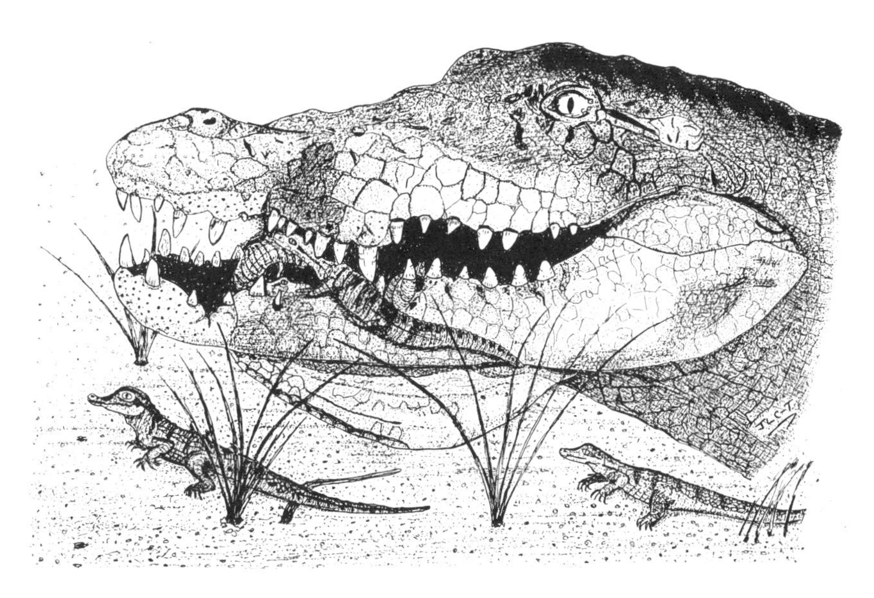

**Fig. 72.** Parent Nile crocodile assisting its newly hatched young to the water. (Cloudsley-Thompson 1994)

The high incidence of ovoviviparity, however, implies increased parental investment.

In most aspects of their biology discussed in earlier chapters, amphibians show far less diversity than reptiles. In their reproductive behaviour, however, they are far more diverse. Certain trends appear repeatedly in their reproductive biology. These include: (1) external to internal fertilization; (2) large to small clutches; (3) aquatic to terrestrial breeding; (4) simple to elaborate mate recognition and courtship patterns; (5) abandonment of the clutch to forms of parental care; (6) oviparity to viviparity; (7) free-living aquatic larval stages to forms of direct development on land. The physiology of reproduction and courtship patterns among amphibians, as well as the significance of the breeding season, have been reviewed by Salthe and Mecham (1974). Other reviews are those of Lofts (1974b), Wake (1977), Duellman and Trueb (1986) and Stebbins and Cohen (1995). Reproductive physiology in reptiles has been reviewed by Bellairs (1969) and by Heatwole and Taylor (1987), among others.

# 9 Daily and Seasonal Cycles, Hibernation, Aestivation and Migration

Most living organisms possess innate periodicities which are synchronized with the daily, lunar and seasonal changes that take place in the environment around them. Endogenous or innate rhythms are of ecological importance in that they ensure the preadaptation of plants and animals to the fluctuations of environmental factors. Amphibians and reptiles are ectothermal, as we shall see in the following chapter, and achieve thermal control by behavioural rather than physiological mechanisms. Day-active species, which include the majority of reptiles, are able to achieve a high degree of thermoregulation by moving into the sun or shade, increasing or decreasing bodily contact with the substrate, entering burrows, and so on. Extant amphibians, unlike reptiles, tend to have moist skins and lose water by evaporation at a much higher rate when they come out into the open. It is not surprising, therefore, that they show a greater tendency towards nocturnal habits, especially in temperate regions.

## 9.1 Biological Clocks

Whether an animal be day-active or nocturnal, it will benefit from the possession of a biological 'clock' that informs it when the time has come for it to emerge from its burrow or hiding place – from which it may be unable to see the daylight appearing or fading. Moreover, in many species there is a marked physiological specialization for a nocturnal or day-active mode of life. Diurnal animals have become adapted to withstand relatively higher temperatures and rates of evaporation, bright light and decreased conductivity of the air for odours. Conversely, nocturnal species are adapted to decreased temperatures, higher humidity and increased conductivity of the air for odours. Diurnal animals are exposed more to predators that hunt by sight, nocturnal forms more to predators that use the senses of hearing and smell.

For this reason, the former may need to be more cryptic and better camouflaged than their nocturnal relatives. At the same time, the sensory physiology of day-active predators differs from that of nocturnal ones along the same lines.

The daily rhythms of plants and animals are reflections of their endogenous clocks and tend to be extremely regular. Some animals are active only during certain periods of the day or night, while others exhibit varying kinds of activity at different times. The locomotory activities of West Indian fossorial amphisbaenians and worm snakes, for instance, show evidence of strong endogenous circadian rhythms under experimental conditions in continuous light. In addition to avoidance of high daytime temperatures, it is believed that nocturnalism may also be associated with the activity of the prey, such as burrowing invertebrates and army ants, which are likewise nocturnal (Thomas and Thomas 1978).

The term 'circadian', mentioned above, requires definition. Biological clocks are regarded as self-sustained oscillations whose phase can be reset or entrained by an external synchronizer or *Zeitgeber* (time giver), a term proposed by Jürgen Aschoff in 1954. In their natural environment, most plants and animals display diurnal or nycthemeral rhythms which persist for a period of a day and a night, and are entrained to a frequency of 24h by the daily cycles of light and darkness engendered by the rotation of the earth on its axis. Although these rhythms frequently persist under constant laboratory conditions, the free-running periods are usually either slightly longer or somewhat shorter that 24h. For this reason, Franz Halberg of Minnesota coined the word 'circadian' (from the Latin *circa*, meaning about, and *dies*, a day) to describe them. Persistent lunar and tidal rhythms are likewise called 'circalunadian' (about a lunar day) or 'circalunar', because the period of the bimodal lunar-day rhythm is usually longer or shorter than the period displayed in nature, while endogenous yearly rhythms are called 'circannual' for similar reasons.

The system is not so complicated as it may sound. Most living organisms, including human beings, are able to tell whether it is day or night, whether it is dawn with the sun near the horizon or midday with the sun overhead. However, this does not obviate the advantages of having a biological clock. (It is additionally useful to human beings to have clocks and watches so that they can tell the time after nightfall or during the day when the sun is obscured by cloud.) It is useful to a nocturnal toad or gecko to know exactly when it is safe to leave its daytime retreat and when its prey is likely to be around.

Endogenous rhythms seldom exhibit their innate periodicities under natural conditions because these are synchronized by environmental changes, just as we synchronize our watches by radio time signals. In general, the activity rhythms of nocturnal animals tend to be delayed by constant light, while those of day-active animals are accelerated (Fig. 73). This is known as the 'circadian rule'. Although a number of exceptions are known, it is significant that many of them are provided by tropical species in whose natural environments the length of daylight varies little throughout the year. When it does occur, however, this shift in phase allows the daily rhythm of an animal to keep pace with the seasons as the days lengthen in spring or draw in during the autumn, for the duration of daylight in temperate regions varies significantly according to the season. Obviously, if an animal is to maintain its regular daily activity, it cannot synchronize its biological clock both to dawn and to dusk, since the period between them is variable. It seems, however, that most nocturnal animals tend to use dusk as the synchronizer, while day-active species use the dawn. Not surprisingly, light intensity is the most usual

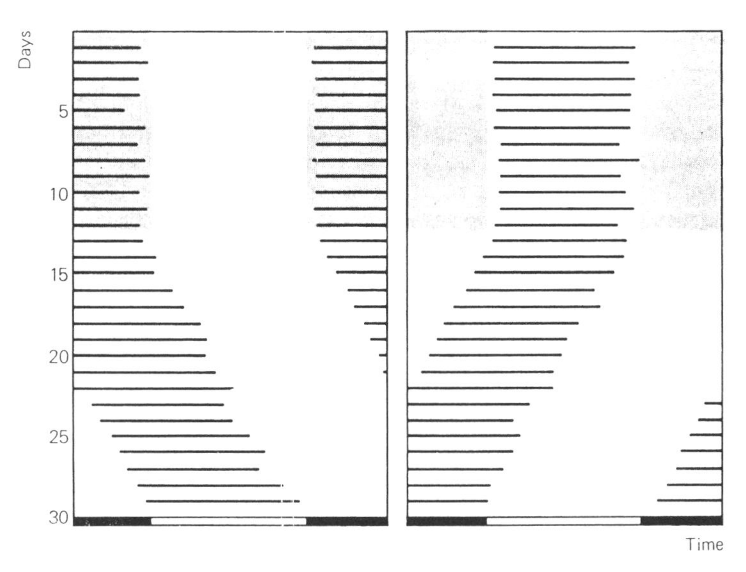


**Fig. 73.** The times of locomotory activity of a generalized nocturnal animal (e.g. a toad) (*left*) and a generalized day-active animal (e.g. a lizard) (*right*) during a 12-day sojourn in a 24-h light–dark regime, followed by 17 days in constant illumination during which the organism typically adapts a free-running period either longer (*left*) or shorter (*right*) than 24 h. *Shaded areas* indicate darkness. *Horizontal lines* show times of activity. *White strip* indicates light period from 0600 to 1800 h. (Cloudsley-Thompson 1980a)

and important synchronizer of circadian rhythms because it is the most regular of environmental climatic variables; but regular temperature changes too may occasionally be effective. In nature, thermal fluctuations (but not, as explained below, absolute temperatures) probably reinforce or supplement changes in light intensity.

Although biological phenomena are usually extremely sensitive to thermal influences, the periods between the peaks of rhythmic activity are relatively independent of changes in the ambient temperature. If this were not so, of curse, biological clocks could only function properly in constant environments. Nevertheless, although the length of the period of the rhythm may be relatively temperature-independent (above the minimum threshold below which activity is suppressed), its amplitude will vary according to the biological temperature coefficient of the process being measured. I shall not attempt to discuss the physiology of biological rhythms any further in the present book, but it should be noted that they play a most important part in the ecology of plants and animals which not only respond to evironmental changes but also are actually able to anticipate them. Biological clocks are responsible for measuring the relative length of daylight; and many cyclical activities, such as breeding and migration, are seasonal and determined by photoperiodism or changing daylength.

The subject of biological rhythms in plants and animals has been discussed in greater detail in my book *Biological Clocks. Their Functions in Nature* (1980), but not with any particular reference to amphibians and reptiles.

## 9.2 Daily Cycles of Activity

Biological clocks are responsible for circadian rhythms of locomotory activity in many amphibians and reptiles. For instance, toads are normally active at night and secrete themselves in cool, damp microhabitats during the day. Under experimental conditions, the majority of individuals of *Bufo americanus* and *B. fowleri* are nocturnal, with a major peak of activity occurring just after the beginning of the dark period. The rhythm is endogenous and persists for weeks in constant darkness. With an increase of 10 °C there is a doubling or trebling in the amount of activity, but no alteration in the frequency of the periodicity (Higginbotham 1939).

Again, the African toad (*Bufo regularis*) is markedly nocturnal, as recorded by aktograph apparatus (a rectangular arena pivoted on a knife-edge about its median axis). Any movement of the toad within tipped the arena and was recorded by a gymbal lever writing on a smoked barograph drum acting as a miniature kymograph. The entire apparatus could, at will, be placed in a refrigerator, an incubator or be exposed to natural daylight and darkness. Modern aktographs use more sophisticated techniques such as infra-red light beams, time-lapse cinematography and automatic counters. Some typical examples of the results obtained with *B. regularis* are reproduced in Fig. 74. From this it can be seen not only that, under natural conditions, the African toad is nocturnal in habit but also that the rhythm persists in constant darkness and constant light – although it gets out of phase with the natural cycle of day and night. The circadian rule is indicated extremely slightly, if indeed it occurs at all. This mechanism, as already mentioned, is normally absent from tropical animals which inhabit regions where there is little annual variation in the length of daylight. Consequently, endogenous clocks there can be synchronized daily at dusk or dawn without the necessity for seasonal shifts in the rhythm (Cloudsley-Thompson 1967, 1980a). In contrast to *B. regularis*,

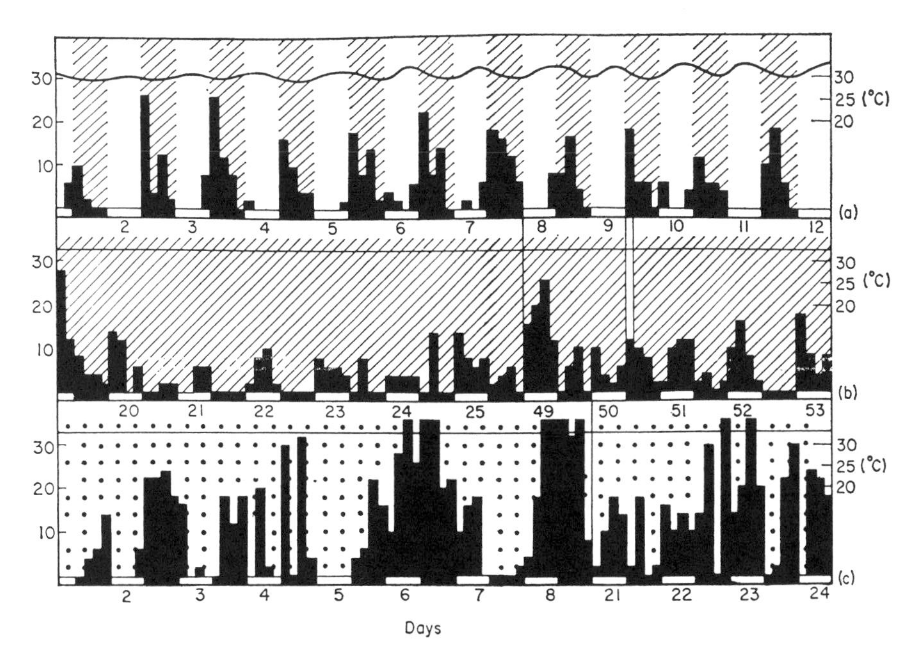


**Fig. 74.** Diurnal rhythm of *Bufo regularis* in moist air. *Black strips* represent 12-h periods from 1800 to 0600 h. *Above* Periodicity in natural daylight and darkness: fluctuating temperature. *Centre* Periodicity in constant darkness and temperature. *Below* Periodicity in constant artificial light and temperature. (Cloudsley-Thompson 1967. Reproduced by permission of the Zoological Society of London)

*Bufo marinus* is preponderantly day-active under natural conditions in the rainforest of Panama, while the European tree frog *Hyla arborea* shows a bimodal pattern of activity, with one peak at dawn and another at dusk.

It might well be supposed that larval Amphibia, being aquatic, would tend to be arhythmic whereas adults would be active at night when water loss by evaporation is reduced. In fact, the reverse is true for *Triturus* spp., the adults of which are arhythmic although the larvae possess a well-developed endogenous nocturnal pattern of activity. This, however, is probably exceptional.

Circadian activity rhythms such as these are invariably accompanied by physiological rhythms, one of the most important of which in the case of both amphibians and reptiles is the daily rhythm of the chromatophores. The adaptive significance of concealment has been outlined in Chapter 6, and it is obviously vital for a vulnerable animal to assume appropriate coloration before it emerges from its hiding place – even though minor adjustments may be made after it has assessed visually the exact hues of its surroundings.

The presence of yellow oil droplets in the eyes of frogs may well be correlated with the fact that frogs tend to be less rhythmic than toads and usually exhibit more daytime activity. The suggestion has been made that the early Amphibia, the Stegocephalia, may also have had yellow oil droplets and been day-active. In becoming respectively more nocturnal and secretive, toads and salamanders have lost the yellow droplets in their eyes.

It is well known that turtles are normally day-active: the retinae of their eyes are composed almost entirely of cones with either slow and slight retinal migrations or none at all. This limits them to diurnal vision and makes them dependent upon olfaction when they are in muddy water. Red oil droplets enable marine species to see through the surface glare of tropical seas. Despite this, however, the majority of chelonian species lay their eggs at night.

The African desert grooved tortoise (*Geochelone sulcata*) is also day-active and shows a well-marked temperature-independent circadian rhythm of activity. Under experimental conditions, this is neither accelerated nor retarded by constant light or darkness. Locomotion is stimulated by light, however, and tends to be reduced in darkness (Fig. 75; Cloudsley-Thompson 1970a). The fact that the activity rhythm of this species does not conform to the circadian rule could again be related to the tropical distribution of the tortoise. Gourley (1972) likewise noted an endogenous activity rhythm in the North American gopher tortoise (*Gopherus polyphemus*). Under experimen-

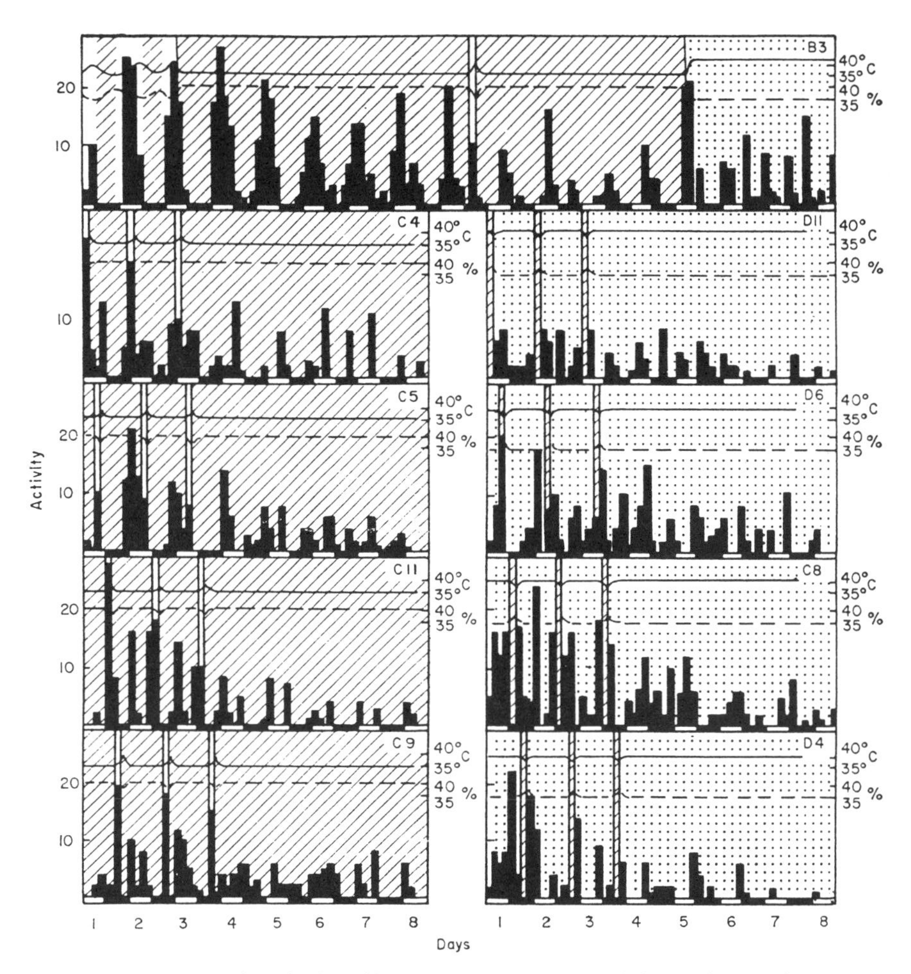


**Fig. 75.** *Top row* Circadian rhythm of locomotory activity in *Geochelona sulcata*, at first in room conditions, then in darkness for several days, apart from one 3-h period of light, and finally in constant light; *lower rows* effect of 3-h periods of light on the activity rhythm of *G. sulcata* kept otherwise in darkness (*left*) and effect of 3-h periods of darkness on the rhythm of tortoises kept otherwise in constant light (*right*). *Ordinates*: activity (*left*); temperature (*solid line*) and relative humidity (*dashed line*) (*right*). *Abscissa*: time in days. *Black strips* represent 12-h periods from 1800 to 0600 h. Although light stimulated activity and darkness inhibited it, neither of them reset the rhythm. (Cloudsley-Thompson 1970a. Reproduced by permission of the Zoological Society of London)

tal conditions of constant light and temperature (26 °C), the biological clock of this species shows a progressive shift in the onset of activity in the direction to be expected for a diurnal animal – each successive period beginning slightly earlier.

Reptiles range from being completely nocturnal to completely day-active, with various intermediate stages between. In some cases, activity patterns may differ between different subspecies, and even among different populations of the same species or subspecies. The time of activity may also vary according to season. Some lizards, especially those that inhabit desert regions where climatic conditions fluctuate wildly, may be nocturnal during summer and day-active in spring and autumn.

Australian elapid snakes seem to have an unusually high degree of diurnality, probably because they depend primarily upon day-active prey. Circadian rhythms are influenced by a complexity of both environmental and physiological variables. Bimodal activity not only may be a direct response to immediate environmental conditions, but also may involve acclimation to temperature. For instance, Graham and Hutchison (1979) found that painted turtles (*Chrysemis picta*), spotted turtles (*Clemmys guttata*) and stinkpots (*Sternotherus odoratus*) acclimated at 15 °C had a unimodal rhythm. When acclimated at 25 °C however, the rhythm was bimodal. Under fluctuating temperatures between these extremes, the rhythm of both painted and spotted turtles was unimodal, with a major pre-noon peak and an afternoon minimum coinciding with the peak of the temperature cycle, whereas the rhythm of the stinkpots became bimodal. Invariably, however, diurnal activity was greatest under long photoperiods, and nocturnal activity greatest when the photoperiod was short.

Under semi-natural conditions, Nile crocodiles exhibit a weak diurnal rhythm of activity with a peak in the early hours of the evening. The rhythm is less marked in hot than in cooler weather and the animals show a greater tendency to haul out of water and bask in the sunshine on days when the air temperature is about 24 °C than in cooler or hotter weather. They also show a tendency to wander far from water at night (Cloudsley-Thompson 1964). Under experimental conditions, young crocodiles usually pass the day on land and spend the night in water (Cloudsley-Thompson 1969).

Many day-active terrestrial reptiles emerge from their burrows or retreats at the same time each morning and retire about the same time each night, whatever the ambient temperature. Other species emerge and retire at specific environmental temperatures so that the times may vary from day to day or season to season. Temperature-dependent emergence may occur at one season and temperature-independent emergence at another. This has been demonstrated by Heatwole (1970) in the Australian desert dragon (*Amphibolurus inermis*). During the summer, this emerges at sunrise each morning and retires at sunset each evening. In winter, however, the animal may be active outside its burrow for a short period only around the middle of the day. Heatwole suggested that as long as the burrow temperature remains above the voluntary minimum, desert dragons emerge on a light-dependent or time-dependent schedule; but that activity is inhibited when temperatures drop below the threshold.

In equable tropical climates not all species make use of nocturnal shelters. The rainbow lizard (*Agama agama*), for instance, roosts on the surface of branches and similar places at night (Cloudsley-Thompson 1981). As Heatwole and Taylor (1987) point out, such animals must be alert from dawn to dusk if they are to avoid predators. They maximize the length of the activity cycle, sometimes at the expense of thermoregulatory precision. In contrast, species that make use of burrows and retreats adjust their periods of activity to coincide with the particular times of day or season at which precise thermoregulation can be achieved. Thus they sacrifice the length of their daily period of activity in favour of thermoregulatory precision. In such animals, the pattern of activity may vary greatly from day to day, depending upon the immediate climatic conditions.

The diurnal or 'diel' rhythms of squamate reptiles, like those of chelonians and crocodilians, although basically circadian and endogenous are nevertheless influenced by exogenous climatic factors – especially light intensity and temperature. The former usually acts as the synchronizer, while the latter determines whether or not and to what extent the reptile will respond. Temperature cycles can often synchronize endogenous rhythms in constant light or darkness. This is the case in the aktograph, both with the day-active African skink *Mabuya quinquetaeniata* and with the nocturnal gecko *Tarentola annularis* in Sudan (Fig. 76; Cloudsley-Thompson 1965, 1972). The same is true of *Acanthodactylus schmidti* from Kuwait (Constantinou and Cloudsley-Thompson 1985). In contrast, although temperature influences the level of activity, it has little effect on the pattern of the diel cycle of the Australian skinks *Hemiergis decresiensis* and *Lampropholis guichenoti*. (For further details, see Heatwole 1976 and Heatwole and Taylor 1987. Endogenous rhythms in reptiles have been reviewed by Underwood 1992.)

## 9.3 Seasonal Cycles and Phenology

March is a time of great awakening in northern temperate latitudes. Amphibians begin to lay their eggs, and reptiles emerge from hibernation to bask in

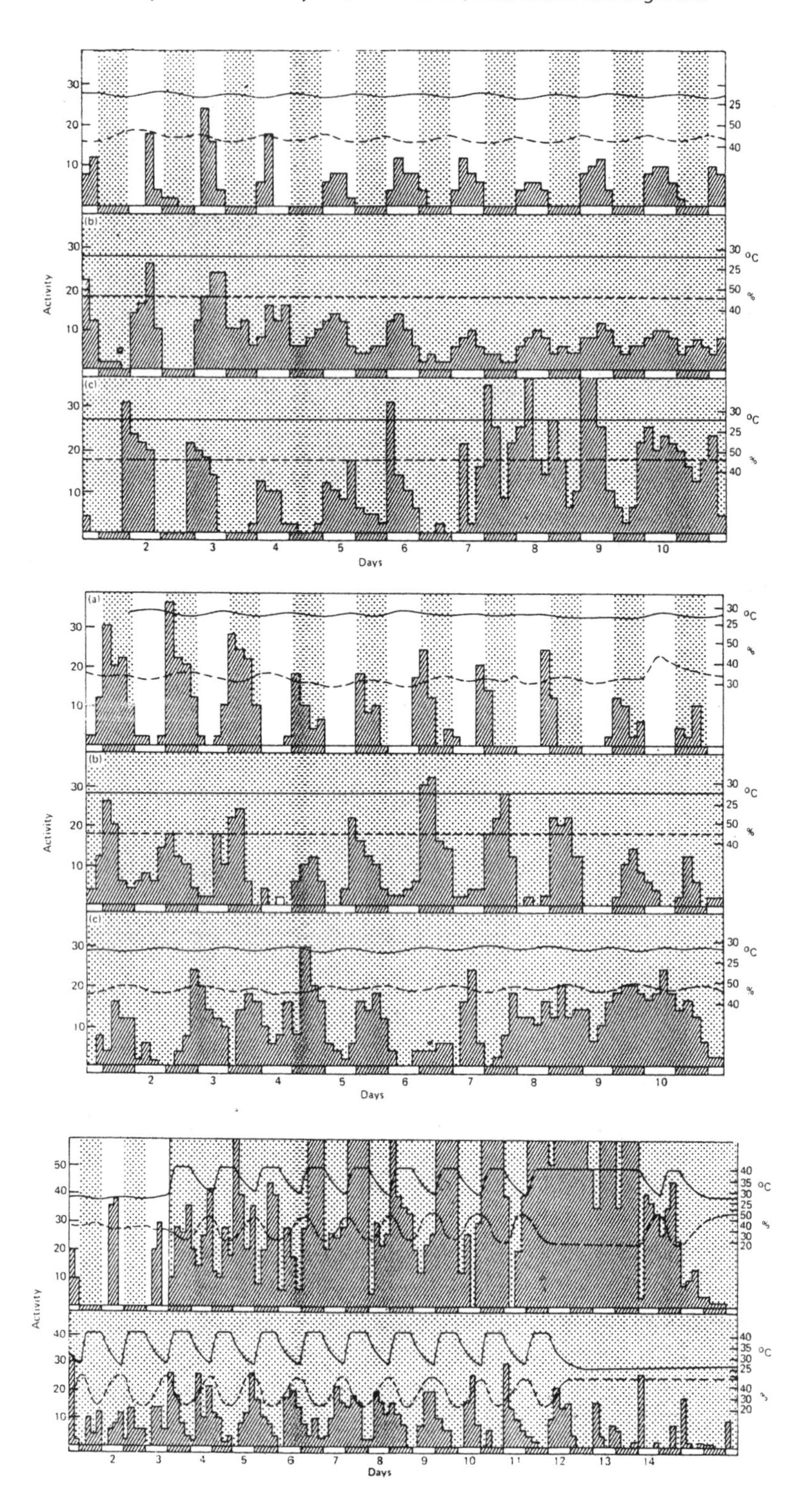
(b)
(c)
Activity
°C
%
Days
(a)
(b)
(c)
Activity
°C
%
Days
Activity
°C
%
Days

the spring sunshine. All amphibians, whether they inhabit temperate or tropical regions, undergo regular endogenous cycles of reproductive activity. Mating, oviposition and larval development are timed to take place when environmental conditions are optimal for these activities. Therefore the gonads or reproductive organs must have been developing in preparation for reproduction for some time before the appropriate environmental conditions arrive. They are, in fact, timed by circannual clocks so as to be absolutely ready to play their part in the reproductive cycle immediately the harsh climate of the winter has ameliorated. Conditions favourable for reproduction may prevail throughout most if not all of the year in damp tropical regions so that breeding can be almost continuous there. In other temperate climates, however, reproductive cycles are distinctly seasonal.

The reproductive organs, testes and ovaries in males and females, respectively, not only produce sperms and eggs, but also secrete the hormones that regulate appropriate behaviour in the two sexes. Amphibian reproduction is controlled by gonadotropin-releasing hormone (GnRH) secreted by the hypothalamus of the pituitary gland, as in other vertebrates. The production of this hormone is correlated with seasonal changes in both Caudata and Anura, but such correlations are highly species-specific. In some instances, the release of GnRH from the hypothalamus is controlled by a negative feedback system, but not in others. Unlike many other vertebrates, amphibians have multiple forms of GnRH. The role of these is not clear: possibly the GnRH receptors on amphibian pituitaries are less specific than their counterparts in other vertebrates. Amphibian reproduction is extremely sensitive to environmental conditions including photoperiod, temperature and humidity (Herman 1992). Photoperiod is measured by the circadian clock, the hypophthalamic cycle by the interaction between this and the circalunadian and the circannual clocks.

**Fig. 76.** Activity cycle of an African skink (*Mabouya quinquetaeniatus*) and an African gecko (*Tarentola annularis*) under various environmental conditions. *Ordinates*: activity (*left*); temperature and relative humidity (in percent) (*right*). *Shaded strips* along the bottom represent 12-h periods from 1800 to 0600 h. *Above* Activity of the skink under (1) normal daylight and darkness, (2) constant darkness, and (3) constant illumination. *Centre* Same as *above* but for the gecko. *Below* Activity of (1) the skink and (2) the gecko in constant darkness and fluctuating temperature. (Heatwole 1976; Heatwole and Taylor 1987, after Cloudsley-Thompson 1965. Reproduced by permission of the Zoological Society of London)

Temperature is thus by no means the only factor regulating the seasonality evident in the physiology of both amphibians and reptiles. There is also evidence of endogenous circannual rhythms in the growth and blood glucose cycles of alligators, and for spermatogenesis in lizards, as well as subsequent reproductive refractoriness and testis regression. Photoperiodism, measured by the circadian clock, may be a proximal factor for some annual cycles, but this is by no means certain. A proximal environmental factor is one that initiates the physiological effects whose ultimate ecological importance may be quite different. For example, increasing daylength is significant only in that it times the development of the gonads.

Although there is considerable interspecific variation in timing mechanisms among reptiles, some general trends are evident. Temperature appears to be a major cue for timing physiological changes in spring after emergence from winter dormancy, while non-thermal mechanisms, photoperiod and endogenous rhythms appear to be primarily responsible for altering physiological activities in the later parts of the cycle (Licht 1972).

In North America, spadefoot toads (*Scaphiopus* spp.) are inactive throughout most of the year. They avoid the harshest desert conditions by aestivating up to a metre below the surface of the soil. Aestivation is really the equivalent of hibernation but takes place in summer. Each year the spadefoot toads emerge on the first night of heavy rain and spawn in the newly formed pools. This usually occurs in early July in Arizona. The toads are stimulated to come up by the vibrations caused by the raindrops, but I think they probably have a circannual clock so that they are physiologically ready to respond to this stimulus. The subsequent period of favourable conditions persists for little more than 8 weeks, and the actual extent of activity is invariably much less. Spadefoot toads are nocturnal during the rainy season, and spend the days buried a few centimetres below ground, emerging to forage only on nights following rainfall when the desert surface is relatively damp. This usually occurs less than 20 times in a season. In September, the toads retreat into their deep hibernation burrows where they remain inactive for the next 9 or 10 months, surviving on reserves of fat accumulated during their brief period of feeding (Tinsley 1990; Tocque et al. 1995). The elusive lifestyle of spadefoot toads imposes severe constraints on the transmission of their parasites.

The life histories of many animals are characterized by the shifting of their ecological niches, and the point in their development at which this occurs can greatly influence their chances of survival. In the tree frog *Agalychnis callidryas*, which lays its eggs in vegetation, there is a trade-off between the risks of predation before and after hatching. When the eggs are attacked by snakes, the tadpoles within hatch rapidly and fall into the water below. Those eggs that are not attacked hatch much later when the newly emerged tadpoles are less vulnerable to aquatic predators. Plasticity in the time of hatching makes it possible for the embryos to respond immediately to the threat of predation on land, but at the cost of increased risk of predation in the water (Warkentin 1995). Early hatching as an escape mechanism does not appear previously to have been described among vertebrates, but a predator-induced delay in hatching has been reported in the salamander *Ambystoma barbouri* where it increases survival and appears not to involve appreciable mortality of feeding costs.

Flatworms (*Phagocotus gracilis*) are well known to prey heavily on hatchling salamander larvae (*A. barbouri* and *A. texanum*), but little on larger more developed larvae. It has only recently been shown by experiment that certain chemicals produced by these flatworms delay the hatching of salamander eggs, whereas chemicals produced by non-predatory isopods do not do so (Sih and Moore 1993). So it is as well for amphibian embryos to be ready or preadapted to hatch, in certain circumstances, before the optimal conditions for their development have been reached.

Like amphibians, reptiles usually breed only at certain fixed times of the year. Those living in northern temperate latitudes tend to mate in spring, and the eggs hatch (or viviparous young are born) in midsummer so that there is sufficient time for them to feed and grow before winter again sets in. An interval between the end of hibernation and subsequent mating may be necessary to allow the eggs to develop sufficiently in the ovaries before fertilization (Bellairs 1969).

Species that would normally reproduce every year do so at longer intervals in the colder parts of their geographical range. Bellairs (1969) pointed out that *Vipera berus* has a biannual reproductive cycle in Finland and northern Sweden, while some species of rattlesnakes likewise probably breed every other year in the northern areas of their distribution in North America. In contrast, subtropical and tropical species show a tendency to reproduce at more frequent intervals. There may be a second breeding season in the autumn in the case of reptiles that inhabit countries around the Mediterranean.

Some chelonians, including the well-known *Terrapene ornata*, lay multiple clutches each season (Legler 1960), while, in southern parts of the United States, females of the skink *Lygosoma laterale* produce three or more clutches of one to seven eggs at intervals of about 5 weeks throughout the breeding season which may last for up to 4 months (Fitch and Greene 1965). Even in regions of tropical rainforest where reproduction continues throughout the year there are not infrequently fluctuations in its intensity. There are probably the result of circannual rhythms which can be synchronized by quite small changes in daylength, the latter measured by the circadian clocks of the reptiles concerned. The physiology of reproduction in reptiles has been described in detail by Bellairs (1969), Porter (1972), Heatwole (1976), Heatwole and Taylor (1987), Underwood (1992), Zug (1993) and many others.

## 9.4 Hibernation and Aestivation

Hibernation is a means by which many animals, including amphibians and reptiles, are able to survive the winter cold in temperate regions. Both aquatic and terrestrial amphibian hibernators must be able also to withstand prolonged starvation.

Amphibians outside the tropics frequently hibernate during the winter. This permits survival without injury over extended periods of time during which it would be impossible for them to carry on with their normal activities (Porter 1972). Winter torpor is sometimes known as 'brumation' to distinguish it from the hibernation exhibited by endotherms and the simple cold-induced inactivity of ectotherms. This distinction is not universally accepted, however, and, furthermore, hibernation is not physiologically identical in all species of mammals. 'Retrahescence' is a term sometimes given to temporary retreat from adverse conditions. Hibernation occurs in all major taxa of reptiles (Dunham et al. 1988). Aestivation or summer dormancy occurs when environmental conditions are hot and dry. It is analogous to hibernation and normally lasts throughout the length of the dry season, as we have seen in the case of spadefoot toads. Both hibernation and aestivation are under hormonal control. They have been reviewed in detail by Pinder et al. (1992) with respect to amphibians.

Hibernation is a major feature of the annual cycles of reptiles not only in temperate regions but also in the subtropics. It allows them to survive long periods of low temperatures or temperatures that are unsatisfactory for escape from predators or for obtaining food. In arid regions, aestivation may occur in response to seasonal drought. Species in which hibernation is mainly under endogenous (hormonal) control are known as 'obligatory' hibernators, whereas those in which hibernation is triggered by external or exogenous factors are called 'facultative' hibernators. The distinction between the two is not clear, however.

In general, temperature is probably the primary exogenous factor influencing the onset of hibernation in reptiles. The energy reserves used during hibernation are provided mainly by fats or lipids, but metabolism is reduced in species that hibernate at low temperatures. Sometimes more stored lipids are used just before and just after hibernation than during hibernation itself. This is reflected in seasonal variations in the components of the body. Males usually emerge before females and thereby presumably increase their chances of encountering receptive females. Successful reproduction may be an important benefit accruing to communal hibernators, especially in extreme environments where the growing seasons are short. A long hibernating season means a short period for feeding, growth and reproduction. Consequently, the reproductive cycles of female ovoviviparous snakes may be extended to several years. (This subject has been reviewed by Gregory 1982.)

Aestivation in reptiles shows many parallels with hibernation. It may be continuous, as in *Varanus exanthematicus*, or discontinuous, as in *Gopherus berlandieri*. These two examples are probably extremes of a continuum. Again, as in hibernation, aestivation may be obligatory in some species (e.g. *V. exanthematicus*) and facultative in others (e.g. *V. niloticus*). The factors that stimulate the onset of aestivation are not known. Temperature or drought may play a part, but it seems more probable that aestivation may be a response to the increasing dehydration of the food, because further intake of dry food would increase the problems of desiccation. Although neither starvation nor dehydration is sufficient to stimulate dormancy in the chuckwalla *Sauromalus obesus*, metabolism is reduced as a response to dehydration in all seasons, and especially at high temperatures (see Gregory 1982).

In conclusion, reptilian reproductive cycles are usually closely coordinated with environmental factors. Seasonal cycles of reproductive activity are characteristic of the reptiles of temperate regions, while the reproduction of tropical forms is frequently acyclic or continuous as in amphibians of the moist tropics, although some species also exhibit cyclical reproductive patterns. Most of the reptiles of temperate regions breed during the spring, and may be stimulated to reproductive activity by the temperature increase that they experience when they emerge from hibernation. Adders (*Vipera berus*), garter snakes (*Thamnophis* spp.) and rattlesnakes (*Crotalus* spp.) have been reported hibernating together in large numbers in communal dens – in caves or down large holes in the ground.

## 9.5 Dispersal and Migration

After hibernation, most adult amphibians in temperate regions migrate to ponds and streams where they congregate and mate. In the amphibians that undergo seasonal reproductive migrations, not only do visual and olfactory cues play a role, but also gravity is used for orientation in regions where the topography is abrupt and breeding pools lie in the bottoms of valleys. During their migrations, salamanders and frogs may move considerable distances, sometimes up to 3 km, and return annually to the same breeding place. Amphibious and aquatic species often select a section of shoreline where they engage in reproductive choruses and mating. Centres of activity such as these provide a diversity of food and easy access to the land for the young after their metamorphosis. There is evidence that individual animals become familiar with their chosen shoreline and that their movements are oriented to it. Actual measurements of home ranges and distances migrated are few in number and not always reliable. There is scope here for considerable research. Toe clipping, a common method of marking amphibians and reptiles, can be used so that individuals may be recognized when they have been located. Another method is to shoot coloured fluorescent pigments into the dermis with a small air rifle. This apparently causes less stress to the animals than toe clipping, and the recapture rates are higher (Nishikawa and Service 1988). Although radioactive tagging yields very high recapture rates, it is time-consuming and expensive.

A very tempting assumption, namely that a breeding chorus attracts members of the species of both sexes to the breeding locality, has never actually been confirmed. Moreover, factors other than voice are dominant in all species studied during the migrating period. Indeed, it is surprising that the calls of many species are so loud that they can be heard at a considerable distance. This may be due to selection by females of males with the loudest voices which are best heard in the chorus – but this apparently has not yet been confirmed (Schiøtz 1973).

Dispersal of squamates after breeding appears not to have been studied in much detail, apart from a few analyses of homing behaviour and, to some extent, of site tenacity. The situation is very different with regard to birds and mammals whose dispersal after breeding has attracted considerable attention. A recent study in Sweden by Olsson et al. (1997) showed that the dispersal of female sand lizards (*Lacerta agilis*) after breeding was not determined by age, size, body condition or number of partners. Nor did dispersal over longer distances by either females or males result in increased mortality. However, females with a low reproductive output dispersed further than others that had reproduced more sucessfully, while males whose physical condition was better did not disperse as far as those in less good condition. Closely related males tended to move further than others that were less closely related. It was suggested that by mating with many partners, reproductive success was increased. The heterozygosity of the parents was strongly correlated with the survival of their offspring.

Sea snakes are believed to migrate into shallow tropical waters during the stormy monsoon season and to return to deeper water during the dry season. Such movements, towards and away from the coast, do not involve travel lover great distances (Cogger 1975). The large breeding aggregations of *Laticauda* spp. reported in the Philippines and of *Astrotia stokessii* in Indonesian waters may involve thousands of snakes travelling to localities particularly favourable for reproduction. Nothing is known, however, of the extent of their journeys. The yellow-bellied sea snake (*Pelamis platurus*) of the eastern Pacific drifts passively most of the time at the mercy of surface water movements. On days when there is little wind, slicks are formed off the coasts of Mexico and Central America which often contain hundreds or even thousands of snakes. On windy or choppy days, in contrast, sea snakes are widely dispersed and, when stranded on tropical beaches, may be unable to crawl back into water. Probably no other vertebrates of the same size are so passively planktonic as yellow-bellied sea snakes (Dunson 1975b). In contrast, *Laticauda* spp. have shown surprising ability to cross deepwater barriers and establish populations in island groups such as those of Fiji and Tonga (Minton 1975).

The strictly tropical distribution of sea snakes is primarily a result of their sensitivity to temperature. Yellow-bellied sea snakes are rarely found in areas where there is a single month in which the average temperature of the sea surface lies below 20 °C. This is true even of equatorial regions such as the Galapagos Islands and Peruvian coast where the sea is relatively cool. On the warmer Equadorian coast, on the other hand, sea snakes are extremely numerous. At the same time, *P. platurus* has an upper lethal temperature of about 33 °C. Since the surface water of the tropical seas in which these snakes live is often as high as 31 °C or even more, they could perhaps have a problem in keeping cool; but they probably swim down to cooler waters and then return to the surface – just as terrestrial lizards shuttle between sun and shade (Dunson 1975b).

A positive correlation probably exists between body size and the home ranges of lizards and snakes, but the data to test it have not yet been collected. Too many variables affect the patterns of movement of squamates, while too few studies have been published in which these variables have been measured to allow any general conclusions to be reached (Macartney et al. 1988).

With few exceptions, crocodilians and squamates do not disperse or migrate great distances, although the range of the estuarine crocodile (*Crocodylus porosus*) is exceptionally extensive and these sea-going crocodiles often travel for considerable distances. The only long-distance migrators with true powers of navigation are sea turtles. The green turtle (*Chelonia mydas*) not only crosses vast distances, up to 2000 km, but also is probably the species that has been studied the most. Nesting used to take place on particular sandy beaches throughout the tropical and subtropical regions of the world where the average temperature of the surface water during the coldest months of the year does not fall below 20 °C (70 °F). Only two of these nesting beaches now remain – a small area of the Caribbean coast of Costa Rica and Ascension Island in the Atltantic Ocean. Mating occurs in the water, just off the nesting grounds, and the males never go ashore.

As we saw in Chapter 7, green turtles swim these long distances to the turtle grasses or seaweeds on which they feed. Once mature, they perform a regular migratory circuit which takes 2–3 years before they return to their nesting home ranges. The migratory circuits and degree of return of males are totally unknown. Other genera of sea turtles include *Caretta*, *Eretmochelys*, *Lepidochelys* and *Dermochelys*. The loggerheads (*Caretta caretta*) and related species of tropical, subtropical parts of the Atlantic, Pacific and Indian oceans also undertake fairly long migrations. Dunham et al. (1988) describe practical methods for marking reptiles in the field.

## 9.6 Navigation

Amphibian migrations seldom exceed a few kilometres, although these comparatively sluggish animals tend to make their way to precisely defined places.

In the case of frogs and toads, navigation by the sense of smell may be assisted by sound and sight, and there is evidence that time-compensated celestial orientation is also used. One of the functions of the circadian clock is to measure time so that allowance can be made for the apparent movements of the sun or the stars across the sky. These invoke sun-compass or celestial navigation.

Both magnetic and time-compensated celestial 'compasses' have been demonstrated in amphibians and reptiles, as well as in birds. The use of a magnetic compass in homing has also been demonstrated in newts and salamanders. For example, specimens of *Notophthalmus viridescens* collected in the field were housed in aquaria oriented in a north–south direction. When tested in a dry enclosed arena and exposed to four magnetic factors, north, east, west and south, they showed a weak orientation along the axis of the holding tank during most of the year. At the time of the spring migratory period (April–early May), however, the bimodal response shifted to coincide with the orientation of the pond from which the newts had been collected. Much stronger orientation was elicited by elevating the temperature of the water in the aquaria to 31–34 °C immediately before testing the newts in the dry arena. They also oriented towards an artificial shoreline. Field experiments show that newts respond to increased water temperatures by leaving the water and sheltering in cooler microhabitats on land (Phillips 1987).

Sun-compass orientation, supplemented by odour gradients in the sea, may play a part in the long-distance navigation of marine turtles. Experimental attempts to demonstrate solar navigation in box turtles (*Terrapene carolina*) and lizards have also produced positive results. More recently, several populations of sea turtles have been found to navigate by responding to a unique combination of magnetic field intensity and field line inclination. It has been shown experimentally that hatchling loggerhead turtles can distinguish between magnetic and inclination angles. They can also distinguish between different field intensities along their migratory routes. Thus they possess the minimal sensory abilities necessary to approximate their global postion using a bicoordinate magnetic 'map' (Lohmann and Lohmann 1996).

Navigational systems have developed along similar lines in amphibians, reptiles and birds. Salamanders, toads, turtles, alligators and birds have all been shown to return to their origins of displacement in the absence of familiar landmarks or goal-emanating cues. In the case of alligators, recent experiments indicate that yearling *Alligator mississippiensis* rely on route-based information while older juveniles have developed a site-based navigational system and are thus insensitive to outward journey effects. Navigational ability improves with age. Yearling alligators are more subject to disruptions of their homing ability by the conditions during their displacement – such as whether the container in which they are being transplanted is open or closed, and the mode of transportation (fast-moving car vs. slow-moving boat) – than are older juveniles (Rodda and Phillips 1992).

Opinions as to the mechanisms and sensory physiology of animal migration are still divided. In many ways, amphibians and reptiles provide excellent material on which to investigate how navigation can be achieved. Every new discovery, however, raises new hypotheses and further questions. Only a few decades ago this phenomenon was shrouded in mystery. Although much remains to be discovered, the hypotheses of today are at least firmly based on observation and experiment.

# 10 Thermal Diversity and Temperature Regulation

Temperature is one of the most pervasive and significant of environmental factors as far as amphibians and reptiles are concerned. It affects almost all their physiological and biochemical processes, including those that influence behaviour. Amphibians and reptiles include among their number species that exhibit varying degrees of thermoregulation and, because of their phylogenetic position, are of crucial importance in the study of the evolution of thermal control mechanisms.

## 10.1 Terminology

Few fields of animal physiology are as beset with semantic confusion and misunderstanding as thermoregulation. For this reason, it will be necessary to explain what is meant by the various terms that are customarily in use today. At one time, both amphibians and reptiles were said to be 'cold-blooded' in contrast to warm-blooded birds and mammals. When it was realized that a crocodile sunning itself on a sand bank might be warmer than a dik-dik in the undergrowth nearby, however, it was realized that the appellation cold-blooded was inappropriate. It was therefore replaced by the term 'poikilothermic', implying that wide variations in body temperature are experienced as a result of changing environmental conditions – in contrast to 'homoiothermic' or 'homeothermic' for animals that maintain relatively constant ($\pm$ 2 °C) body temperatures. More recently, the terms 'ectothermic' and 'endothermic' have been used more frequently. Ectothermic means that the primary source of warmth comes from the environment. In contrast, endothermic implies thermoregulation based primarily on metabolic heat production. Care should be used in the application of both these terms, however, because most terrestrial animals use both modes of heat gain to varying degrees. The same is true of the terms 'bradymetabolic' to describe the relatively low metabolic

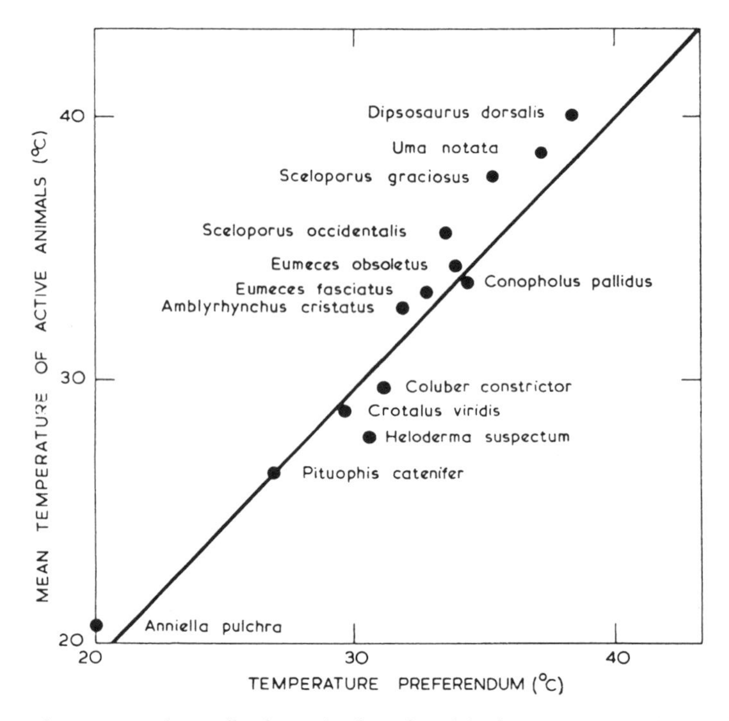


**Fig. 77.** Experimentally determined preferred body temperatures in reptiles in relation to temperatures of active animals observed in the field (based on the work of several authors). (Cloudsley-Thompson 1971)

rates of ectotherms and 'tachymetabolic' the higher metabolic rates of endotherms.

Most animals respond to thermal gradients by congregating with some degree of precision about a temperature which is characteristic for the species concerned. This is called the preferred or 'eccritic' temperature and is not infrequently influenced by the temperature to which the animals have previously been acclimated. Because many of them have moist skins, the concept of eccritic temperatures is seldom meaningful in the context of Amphibia, but it is more significant for reptiles (Fig. 77; Cloudsley-Thompson 1971). Transpiration is more important in the thermoregulation of amphibians than it is of reptiles.

## 10.2 Size and Thermoregulation

Although the reptilian mode of organization has allowed for the evolution of the largest terrestrial animals ever to have lived – dinosaurs – very small species are found only among lizards. The smallest snakes (*Typhlops* spp.) may reach lengths of 200 mm, the smallest chelonians (*Homopus signatus* on land, *Clemmys muhlenbergii* and *Kinosternon baurii* in freshwater) seldom exceed 100 mm, while the smallest crocodilians (*Palaeosuchus palpebrosus* and *Osteolaemus osborni*) only reach 1.2 m in length. There are no very small amphisbaenians, however, and *Sphenodon* spp. grow to 60 cm. In contrast, several species of lizards have adult body lengths less than 40 mm and weights of less than 1 g. Most of these are geckos (Avery 1996).

Probably the most important thermal aspect of size is the relationship between mass and surface area or, as it is commonly expressed, the surface to volume ratio. As we saw in Chapter 2, every time the surface area is squared, unless the shape changes, the weight of any object is cubed. For this reason, smaller animals heat up and cool down very much more rapidly than do larger animals of the same general form. It is significant that very small lizards, less than 20 mm in length and weighing less than 0.5 g, form less than 1% of about 1200 species and are found only in the tropical West Indies and all small species are restricted to tropical regions. It is probable too, that one of the functions of the 'sails' of the pelycosaurs (Fig. 6) and the bony plates of *Stegosaurus armatus* (Fig. 54) may have been thermoregulatory, as we have already seen. The subject will be discussed in further detail below.

Many physiological processes occur faster in smaller than in larger animals. For example, small amphibians and reptiles heat up more quickly in the sun, and dehyrate more rapidly, than larger ones. In some cases this presents no problems because the tempo of life is faster for smaller species whose hearts beat more quickly and whose gestation periods and life spans are usually shorter than those of larger species. Indeed, these characters may be advantageous. Smaller species are able to exploit smaller, more abundant prey, compete favourably with larger species, and be able to escape from predators in a wider range of crevices and other refuges. At the same time, they probably have a greater range of potential predators, possess poorer vision owing to a reduction in retinal cell size and a denser packing of neurons in the visual cortex of the brain, reduced clutch sizes, and a decrease in stamina and endurance. Size reduction in chelonians and crocodilians may be limited by their long armour; but the question remains as to why the smallest snakes should be larger than the smallest lizards. The answer may lie in the fact that there is a strong correlation between size and the number of vertebrae present. There must be a minimum vertebral number to permit effective serpentine locomotion. A second constraint relates to the problem of supplying energy to an elongated body via a relatively small mouth – although this is largely overcome in snakes by modifications of the skull which permit relatively enormous prey to be ingested (Avery 1996). Moreover, an elongated body increases the surface to volume ratio. Consequently, long and thin reptiles heat up and cool down more rapidly than do squat, heavily built forms.

McNab (1978) suggested that the large changes in body size that occurred during the evolution of the mammal-like reptiles were crucial to the development of endothermy which, indeed, was first achieved by increasing thermal independence through an increase in mass. Then, by modifying the thermal properties of the body surface, mass could have decreased with only moderate reduction in the total rate of metabolism. This would have resulted in small animals that were essentially mammalian in their energetics and thermoregulatory capabilities.

Later, Turner and Tracy (1986) argued that, crucial to this scenario, is the contention that large body size necessarily renders an animal homeothermic. Certainly, when the ambient temperature changes large reptiles change their body temperatures more slowly than small reptiles and are therefore inertial homeotherms. Both homeothermy and high body temperatures are of great physiological and ecological value (Heinrich 1977): many extant lizards thermoregulate and maintain high body temperatures by means of behavioural and physiological adjustments (Huey and Slatkin 1976). Nevertheless, these do not need to be associated. Galapagos iguanas (*Conolophus pallidus*), for instance, regulate their body temperatures in such a way that a constant temperature is maintained throughout the day, although very much higher temperatures could easily be attained. The actual temperature at which their bodies are regulated seems within limits to be of minor importance (Christian et al. 1983).

There are two ways in which the large therapsids of the Late Permian might have achieved homeothermy. If their large size was essential, as McNab (1978) suggested, they might have been inertial homeotherms with rather low body temperatures. In that case, endothermy would have maintained their bodies at relatively low temperatures, with the attainment of high temperatures appearing later during the evolution of the mammals. Alternatively, if heat exchange was controlled in the later mammal-like reptiles, then high body temperatures would have evolved first, with endothermy appearing later.

Turner and Tracy (1986) calculated that the larger and more advanced of the pelycosaurs, such as *Dimetrodon* and *Edaphosaurus* spp., may well have pursued the first method, developing an ancillary heat exchanger in the form of their dorsal 'sail' (see below). During the evolution of the therapsids, however, the limbs were rotated ventrally so that their possessors walked and ran more in the way that mammals do. At the same time, therapsid limbs became more slender and the tail was greatly reduced in size. These changes could well have enhanced heat loss if the animals had been extremely active and maintained high body temperatures.

## 10.3 Behavioral Thermoregulation

Whilst poikilothermy may be a disadvantage to Amphibia in some respects, it nevertheless serves their economy in numerous ways, including enabling them to survive for long periods without food, minimizing radiation heat loss, enabling them to tolerate low oxygen tension and to be much smaller than homeotherms. Amphibians can modulate their body temperature (Tb) by moving among appropriate microhabitats – between water and land, burrows and sunlight, and so on. Basking in the sun is apparently uncommon among urodeles but has been described in several anurans.

The Tb of an amphibian in its natural environment results from complex interactions of numerous environmental factors and internal physiological adjustments. The latter will be mentioned later in this chapter. An extraordinary example of behavioural thermoregulation is presented by canyon tree frogs (*Hyla arenicolor*). During daylight hours these bask on vertical rock faces about a metre above the waterline along permanent waterways of the North American deserts. Snyder and Hammerson (1993) examined their thermoregulatory activity by measuring Tb and evaporative water loss (EWL) in the field, and EWL under controlled laboratory conditions. Around midday in June near Grand Junction, Colorado, Tb (30.6 $\pm$ 0.2 °C) was independent of air temperature, water temperature and that of the surrounding rock surface. Between dawn and dusk, the EWL of basking frogs averaged 24.8 $\pm$ 3.5% of body mass, which is equivalent to the bladder water reserves of the animals (25.1 $\pm$ 1.4% of body mass). The average resistance of the skin to water loss (14.9 $\pm$ 0.3 s/cm) was nearly seven times greater than that of a typical terrestrial frog (*Rana pipiens*: 2.2 $\pm$ 0.3 s/cm) but only about one fifteenth of that of frogs adapted to arid environments. Because *H. arenicolor* inhabits a relatively arid environment, it is surprising that its rate of EWL, although significantly lower than the values for normal frogs, is not closer to that of arid-adapted species. Its unique natural history, however, may limit the adaptive value of a lower rate of EWL since it is able to rehydrate on a daily basis. Furthermore, it has a large bladder which can store up to about 25% of its body mass in dilute urine from which the body fluids can be replenished. The adaptive advantage of thermoregulating by sun basking and simultaneously evaporating is not immediately apparent.

In general, most amphibians on land maintain some degree of thermal homeostasis behaviourally by a combination of sun basking and seeking cooler microhabitats. Behavioural thermoregulation in Amphibia has been reviewed, among others, by Brattstrom (1970), Duellman and Trueb (1986), Hutchison and Dupré (1992), Stebbins and Cohen (1995) and Warburg (1997).

Times of activity of the majority of reptiles are relatively consistent from day to day and change more or less regularly with the seasons. Many diurnal lizard species have a bimodal activity rhythm and are active at dawn and dusk during summer months, but have a single midday peak of activity at cooler times of the year. This presumably facilitates thermoregulation. Sympatric species often differ in their activity rhythms, some emerging later, after others have become less active (Pianka 1977).

Although most geckos are nocturnal, some species feed actively during the day and bask in direct sunshine. Many others are active during the day but do not move out of the shade, while some nocturnal species are occasionally active in sunlight. Of the tropical American geckos, *Thecadactylus rapicaudus* is strictly nocturnal, whereas *Sphaerodactylus lineolatus* is diurnal. In nearly all the nocturnal species that have been investigated, the bulk of nocturnal activity takes place before midnight and before temperatures begin to drop appreciably. Maximum activity in the arboreal Australian *Gehyra variegata*, for instance, takes place during the first 3 h after sunset (Fig. 78; Bustard 1967).

The function of the spectacular dorsal 'sails' of some of the Pelycosauria (Fig. 6) has generated much speculation. They might have been used in intraspecific displays, according to Bakker (1971). At the same time, as already mentioned, the large area of the sail has long been suspected of being a thermoregulatory device for absorbing solar heat early in the morning, and for radiating it later when the animals became overheated (Romer 1948). This hypothesis was supported by the calculations of Bramwell and Fellgett (1973) which showed not only that the sail of *Dimetrodon grandis* could well have enhanced heat gain, but also that it would have been capable of radiating heat.

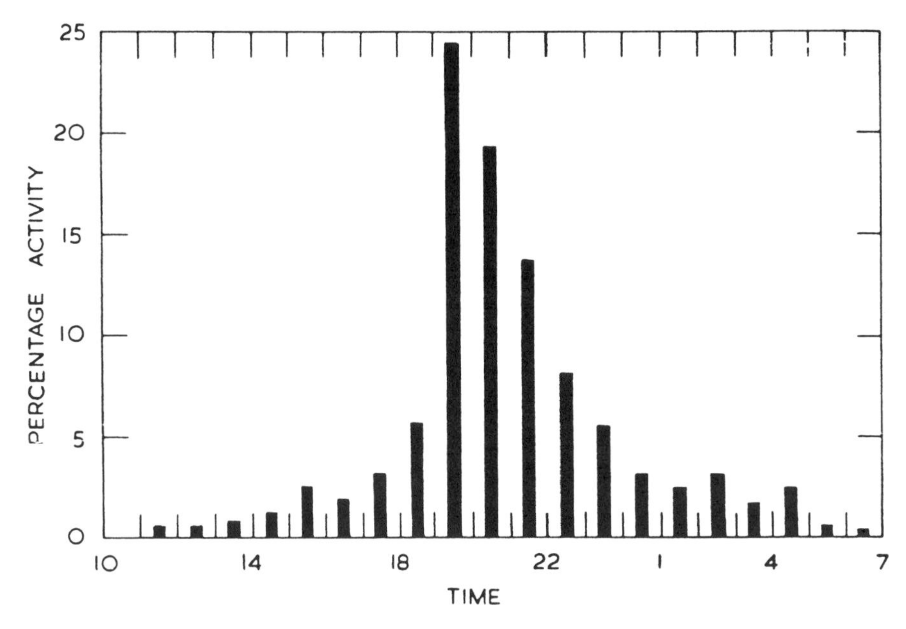


**Fig. 78.** Hourly percentages of total activity of *Gehyra variegata* recorded in an aktograph out-of-doors under normal summer fluctuations of light and temperature. (Ten pairs of animals, 60 records.) Sunset between 1900 and 1920 h. (Cloudsley-Thompson 1971; after Bustard 1967)

Furthermore there is absolutely no reason why it should not have had an additional function of display.

Faster attainment of the activity temperature in the morning would have been advantageous to a carnivorous reptile that fed on large poikilothermic animals while these were still torpid or sluggish. Body size and sail area increased during the evolutionary history of the sailed pelycosaurs. The earliest forms, such as *Dimetrodon milleri* (ca. 50 kg), were too small to have been inertial homeotherms, but *D. grandis* (ca. 250 kg) could well have maintained low body temperatures throughout the day, in which case the sail would have conferred no thermoregulatory advantage. Probably homeothermy and high body temperatures evolved simultaneously, early in the history of the mammal-like reptiles. It should be remembered, however, that the therapsids did not arise from *Dimetrodon*, but from a more primitive, unsailed sphenacodont such as *Haptodus*, as indicated by the fossil record (Tracy et al. 1984).

Many existing reptiles seek shade when they begin to overheat, and then emerge into the sunlight as they cool off (Fig. 79). Repeated alternation

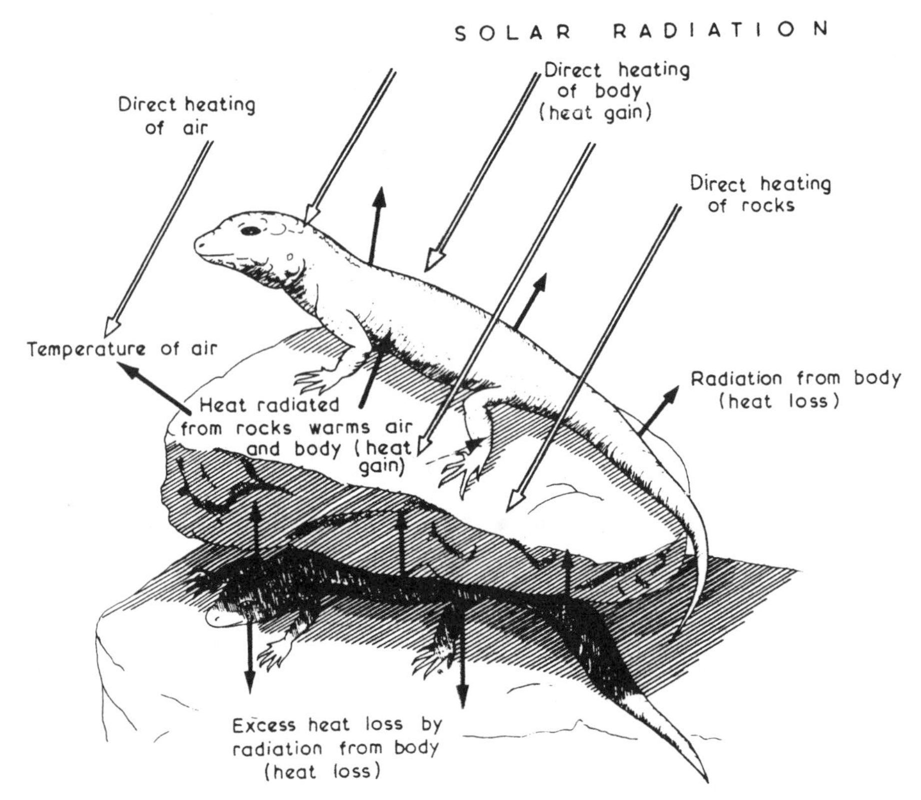


**Fig. 79.** Thermal relationships of lizards in full sun and shade. (Cloudsley-Thompson 1971; after Harris 1964)

between basking in sunlight and seeking shade is known as 'shuttling'. It is a characteristic pattern of behaviour in numerous species and has been discussed by many authors on numerous occasions. Behavioural thermoregulation by orientation of the axis of the body to the rays of the sun has been developed to its highest degree in day-active lizards (Schmidt-Nielson and Dawson 1964), and must have been equally if not more important to dinosaurs and other extinct reptiles.

When temperatures exceed the optimum, lizards such as *Dipsosaurus dorsalis*, *Holbrookia maculata* and *Sator grandaevus* may take refuge among the branches of trees and bushes where the air is cooler than on the ground. When the sand is really hot, the Namibian lacertid *Aporosaura anchiaetae* lifts its body, feet and tail alternately above the surface in a most comical, jerky way. Contact with the substrate may be used both for heating and for cooling the body. When heated by the sun's rays, lizards and snakes often dig their bodies into the cooler layers of the soil, while nocturnal species press their bodies against the surfaces of rocks while these are still warm from the sun. This, too, is why nocturnal reptiles are so frequently squashed by vehicles at night on desert roads. Behavioural thermoregulation has been reviewed on numerous occasions and by many authors (e.g. Cloudsley-Thompson 1971, 1972, 1991; Bradshaw 1986, 1997; Heatwole and Taylor 1987; Templeton 1970; Huey and Slatkin 1976; several authors in Gans and Pough 1982).

## 10.4 Physiological Thermoregulation

Amphibians have the potential for thermoregulation by a variety of physiological mechanisms in addition to behavioural ones. These include: (1) changes in evaporation from body surfaces (see Chap. 11); (2) vasomotor changes in the peripheral circulation; (3) colour change altering the reflectivity of the skin; (4) internal adjustments of the cardiovascular system, such as circulatory shunts and cardiac output; (5) thyroid-induced changes in metabolism, and muscular activity; (6) alteration of central and peripheral neural elements; and (7) tolerance of high or low temperatures. They have been reviewed in detail by Hutchison and Dupré (1992). The present book is concerned more with ecology and behaviour than with physiology. I shall therefore make no attempt to discuss these ecophysiological matters. Nevertheless, valuable information can be obtained by measuring the body temperatures of amphibians and reptiles in the field. This can be done by inserting a micro mercury thermometer or thermocouple into their mouths or cloacas, as appropriate.

It is interesting to note that the thermal tolerance of frogs may be affected by photoperiod. When leopard frogs are subjected to a regime of 16h of light and 8 of darkness (16:8), their thermal tolerance is significantly higher than it is under short (8:16) or moderate (12:12) photoperiods. It may be of ecological significance that the time of highest temperature tolerance falls during late morning or early afternoon - periods of relatively high environmental temperatures (Mahoney and Hutchison 1969). The main thermal adaptations of amphibians inhabiting arid environments lie in their longer survival at high temperatures, thermal preference and critical thermal maximum (CTM) compared with those of mesic amphibians (Warburg 1997).

Most biologists are aware that the dinosaurs may have been endothermal yet they have not considered the possibility that the mammal-like Therapsida and their predecessors might well have been warm-blooded too. Endothermy, the internal production of heat, entails essentially constant high rates of metabolism that are costly in energy. That is why birds and mammals tend to consume at least ten times as much food as the ectothermic amphibians and reptiles of comparable size and weight.

The world of therapsid reptiles, from the Middle Permian to Late Triassic (ca. 240–190 my B.P.), was one of profound geological changes and great diversification of terrestrial animals and plants. Throughout this time, the continents of today were coalesced in a single continent, Pangaea. However, tectonic movements were taking place and mountain ranges, such as the Appalachians of North America and the Urals of western Asia, were created. Despite the fact that land masses were contiguous and the climates increasingly equable, there were great variations between the northern and southern floras. Therapsids underwent great adaptive radiation and were probably the dominant terrestrial vertebrates of the time. All vertebrates, especially reptiles, expanded in numbers and diversity, despite the large-scale extinctions that took place at the end of the Permian. These led to the disappearance of some 75% of Late Permian reptilian genera. Subsequent radiation in the Early to Middle Triassic period was responsible for the great diversity of reptiles early in the Mesozoic. Considering their long dominance and increasingly sophisticated anatomical and physiological systems, the extinction of the therapsids seems somewhat enigmatic. Many factors were undoubtedly involved, as in the case of the dinosaurs. Probably competition from Late Triassic dinosaurs hastened the end of the Therapsida. There was a marked reduction in the size of the later cynodonts, some of whose skulls were only about 7 cm in length, while the early mammals were no longer than that in total length (Ruben 1986)!

Many authors have proposed that mammalian endothermy had its origin among the mammal-like reptiles. Bakker (1971) pointed out that the short, stocky limbs of the earliest Carboniferous reptiles suggest that these animals were not heliothermic but were active in shaded coal-producing forests. The first archosaurs were similar. By mid-Triassic times, more erect, gracile and active reptiles had evolved. These may have enjoyed endogenous heat production which could have exceeded that of the living mammals of today. A fully erect posture was achieved at the close of the Triassic. The dinosaurs, according to Bakker (1971), were both endothermic and homeothermic in the Mesozoic: they were the precursor of the birds. Advanced therapsids may not have been heliophilic but had reduced surface to volume ratios. The first mammals to which they gave rise would have been hairy and maintained high body temperatures throughout the night and could thus have benefited from nocturnal activity. These primitive mammals were small, but would probably have experienced difficulty in losing heat without efficient evaporative cooling. During the hours of daylight they must have been competitively inferior to dinosaurs.

Changes in body size are a feature in the phylogeny of several lineages of mammal-like reptiles. Carboniferous pelycosaurs weighed less than 1 kg, but, by the Early Permian, the pelycosaurs averaged up to 250 kg in weight. In the Triassic, body size declined again. Many authors have proposed that mammalian endothermy origianted in the therapsids, and McNab (1978) suggested that the large changes in body size that occured during the evolution of the group were crucial in the development of endothermic thermoregulation.

According to Paul (1994a), large dinosaurs were apparently *r*-strategy egg layers with reproductive outputs much higher than those of large K-strategy mammals. The populations of these creatures probably consisted mostly of posthatchling juveniles which were no longer dependent upon their parents for survival but were large enough to keep up with the herd. There is a considerable body of evidence to suggest that juvenile dinosaurs would have grown rapidly, at rates comparable with those of birds and mammals. Lactation would not have been necessary for this high rate of growth. Large extant birds can also grow very rapidly, but high minimal and maximal metabolic rates certainly would have been necessary for dinosaurs to do the same. The gap between the growth rates of terrestrial reptiles and mammals increases with increasing size. This fact contradicts the possibility that the growth rates of gigantic ectotherms would have converged with those of giant endotherms. The *r*-strategy of the dinosaurs helps to explain their long period of success, but exacerbates the problem of explaining their final extinction, discussed in Chapter 3.

Very small, 'altricial' or helpless hadrosaur nestilings probably lived in open nests exposed to the weather. In other to survive and grow rapidly, they would have had to be insulated endothermic homeotherms. The small, altricial young of small ornithopod and theropod dinosaurs would have required similar adaptations unless they were brooded by their parents. If the parents did indeed brood their hatchlings, they probably required elevated metabolism and soft insulation to keep their young ones warm. As these grew, however, they may have shifted from an endothermic to a more reptilian type of physiology in response to having a more thermally stable body mass. It could have taken a year or more for the larger species of dinosaurs to attain the higher thermal stabilities of masses of over a tonne, and rapid growth might have continued for decades after this stage in some of the largest dinosaurs (Paul 1994b).

The temperatures at which dinosaur eggs developed cannot be known, nor whether they experienced thermal sex determination, but the temperature at which the eggs are incubated affects the sex of the hatchlings in most extant families of turtles, all crocodilians and a few species of lizards. In the case of turtles, eggs that develop at lower temperatures produce male offspring, those that develop at higher temperatures hatch into females: the situation is reversed in crocodilians and lizards. The middle third of the period of embryonic development is critical for sex determination in chelonians and the thermal 'windows' are narrow. The period of sensitivity is somewhat earlier in crocodilians, but their embryos are slightly more advanced at oviposition than are turtle embryos. The period of sensitivity has not yet been determined in the case of lizards, nor has the exact way in which temperature intervenes in the differentiation of the gonads (see Packard and Packard 1988).

The earliest known bird, *Archaeopteryx lithographica*, resembled the dinosaurs of its time but had feathers which, according to Bakker (1971), would have been necessary for it to maintain a warm body temperature endothermally. It is significant, moreover, that in the radiation of the dinosaurs no really small form appeared, although the 46-cm (18-in.) *Sinosauropteryx prima* appears to have been feathered. At the same time, at least one of the pterosaurs, *Sordes pilosus* (Fig. 80) of the Jurassic period, was hairy. In contrast, even today large mammals such as elephants, rhinos and hippopotamuses are almost hairless. Overheating, not cold, is a problem for them as I believe it to have been for the larger Cretaceous reptiles (see Chap. 3).

**Fig. 80.** Reconstruction of *Sordes pilosus*, the hairy pterosaur of the Jurassic. (Cloudsley-Thompson 1978b; after Ricqlès 1975)

In comparison with the dinosaurs of the Jurassic and Cretaceous, modern reptiles are mostly small and poikilothermic. Their large surface to volume ratio would render homeothermy uneconomic. Furthermore, they are able to evade inclement seasonal weather conditions by hibernating or aestivating in sheltered retreats. Even so, in addition to behavioural thermoregulation they utilize a number of physiological thermoregulatory processes. For instance, as long ago as 1838, P. Lamarre-Picquot claimed that the Indian python (*Python molurus*) coiled about its eggs and produced sensible heat as an aid to incubation. This statement aroused much controversy. Hutchison et al. (1966), however, showed that during the brooding period *P. molurus* can achieve a marked level of physiological thermoregulation by spasmodic contractions of the body musculature with a considerable increase in metabolism and body temperature at ambient temperatures below 33 °C (Fig. 81). Presumably thermoregulation is only economic to pythons while they are incubating their eggs.

Compared with squamates, turtles and tortoises have a much more compact shape and a reduced surface to volume ratio. For this reason one might reasonably predict that terrestrial species would be more likely to suffer from heat stress than from cold. During the 1960s, when I was Professor of Zoology in the University of Khartoum, Sudan, we kept a number of very large *Geochelona sulcata* in the grounds of the Department. In November 1965, one of the females laid 17 eggs of which only one hatched the following June. The Sudanese laboratory staff named the baby Abdel Gadir (or 'servant of God' in Arabic – a common nickname for tortoises in that part of the world). Until a proper enclosure had been constructed for him, I used to put Abdel Gadir on the lawn to graze and, in order not to lose him, confined him beneath a clear plastic microscope cover in the shade of a palm tree.

One day, on the way home to lunch, I remembered that I had forgotten about Abdel Gadir. I dashed back to find that the sun had moved and he was no longer in the shade. Indeed, the poor little creature was extremely hot and frothing all over his head and neck. I took him back to the laboratory and gave him a drink. When jerboas and kangaroo rats get overheated, they salivate onto their throats. This cools them by evaporation. Could tortoises respond in the same way, I wondered? Next morning, Abdel Gadir found himself in an incubator at 45 °C. I removed him every 30 min to be weighed, and measured his body temperature with a thermistor thermometer. During the first hour, his weight decreased only slightly but his temperature rose from 58 to 61.5 °C; at this point he began to salivate and froth at the mouth. His temperature

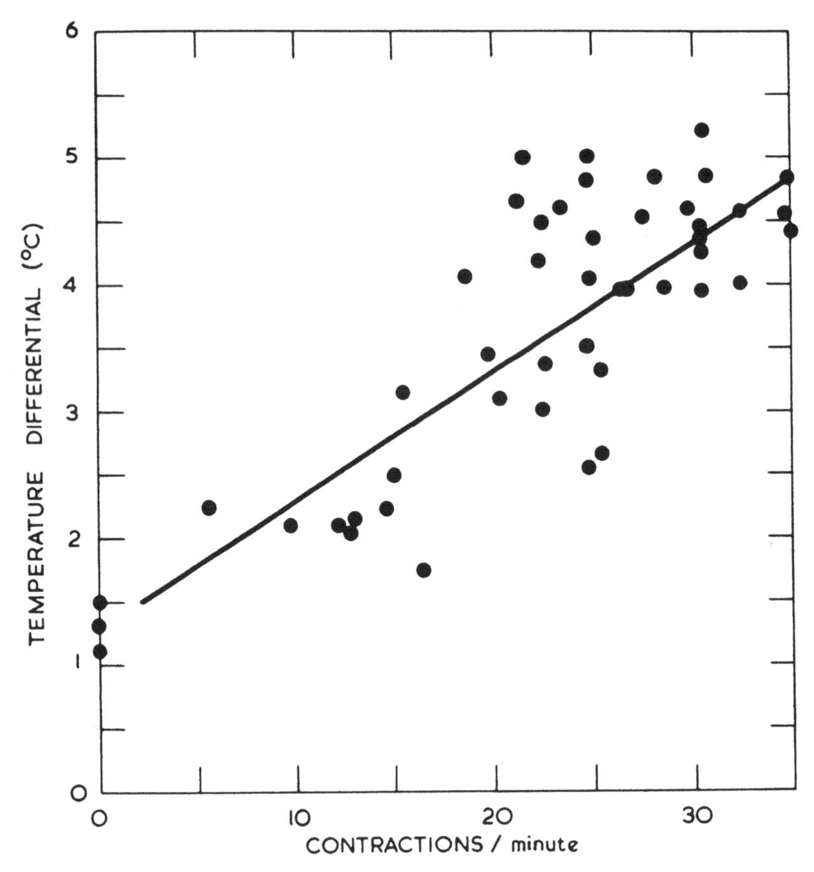


**Fig. 81.** Correlation of rate of contraction with temperature differential (= 0.1 contr./min + 1.3) of body muscles of a brooding *Python molurus*. (Cloudsley-Thompson 1971; after Hutchison et al. 1966)

stabilized, but he began to lose weight rapidly. Subsequently, I obtained many other baby tortoises and was able to repeat the experiment many times (Fig. 82; Cloudsley-Thompson 1970a). In 1969, with my wife and Bud Riedesel, I carried out a similar experiment at the University of New Mexico, Albuquerque, on the box tortoise (*Terrapene ornata*). Not only did this species also salivate when a critical temperature had been reached, but also it urinated on its back legs. We found that the rate of heartbeat increased only when the body temperature increased, but salivation could be induced by heating the head alone (Riedesel et al. 1971). A simlar response occurs in the European *Testudo graeca* (Cloudsley-Thompson 1974). A popular account of these experiments later appeared in *Natural History* (Cloudsley-Thompson 1993).

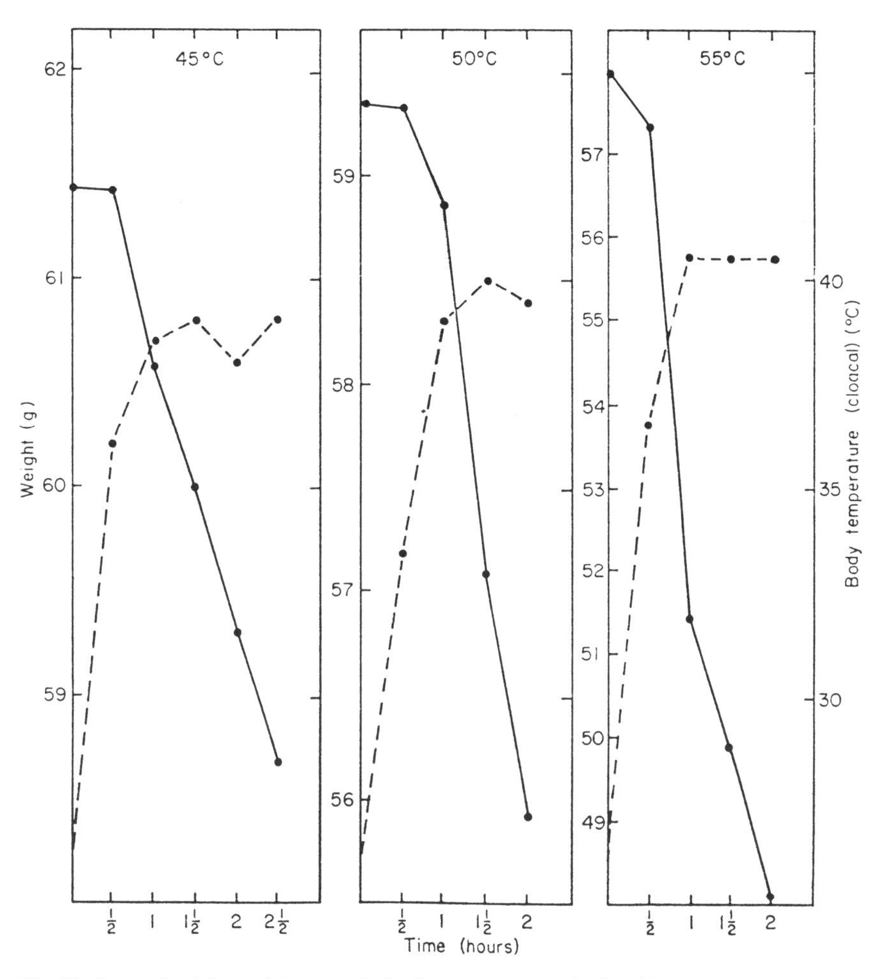


**Fig. 82.** Loss of weight and increase in body temperature of a hatchling *Geochelone sulcata* during consecutive exposures of 30 min in dry air at various high temperatures. Weight (*solid line*); body temperature (*dashed line*). (Cloudsley-Thompson 1970a. Reproduced by permission of the Zoological Society of London)

Studies of the heating rates of horned lizards (*Phrynosoma coronatum*) indicate that emergence from the sand at dawn may be dependent upon the temperature of the head and independent of that of the body. Head temperature in this species is regulated by opening lateral shunt vessels from the cephalic sinuses to the external jugular veins. This by-passes a countercurrent heat exchanger between the internal jugular veins and the internal carotid arteries (Fig. 83; Heath 1966). In consequence, the lizards do not emerge until their heads and brains have warmed up and they are fully alert.

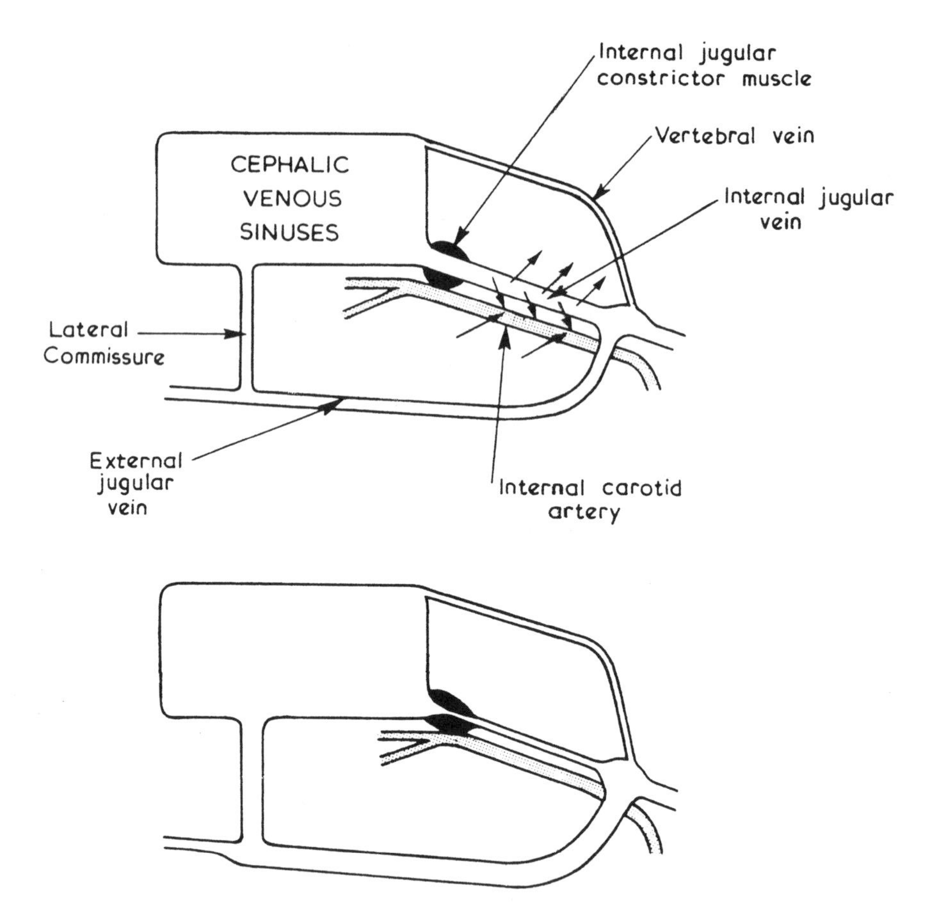


**Fig. 83.** Relations of the major vessels in the head of the horned lizard. The venous spaces of the head are abstracted to a single space. *Above* Internal jugular vein is open. *Arrows* indicate heat exchange between internal carotid artery and tissues of the neck; *Below* closure of the internal jugular constrictor muscle causes collapse of the internal jugular vein. Venous blood returns through a shunt to the external jugular vein. Cool blood, warmed only by the neck tissue, enters the head, while warm blood flows through the external jugular vein to the body. (Cloudsley-Thompson 1971; after Heath 1966)

The subject of jugular shunts and head-body temperatures was discussed in detail by Templeton (1970).

The interactions between behavioural and physiological thermoregulation in reptiles may be extremely complex, as Heath (1965) showed in his elegant research on *P. coronatum* (Fig. 84). At temperatures approaching the maximum voluntary tolerance, shade seeking and burrowing are replaced by panting and cloacal discharge. Again, the primary advantage of basking

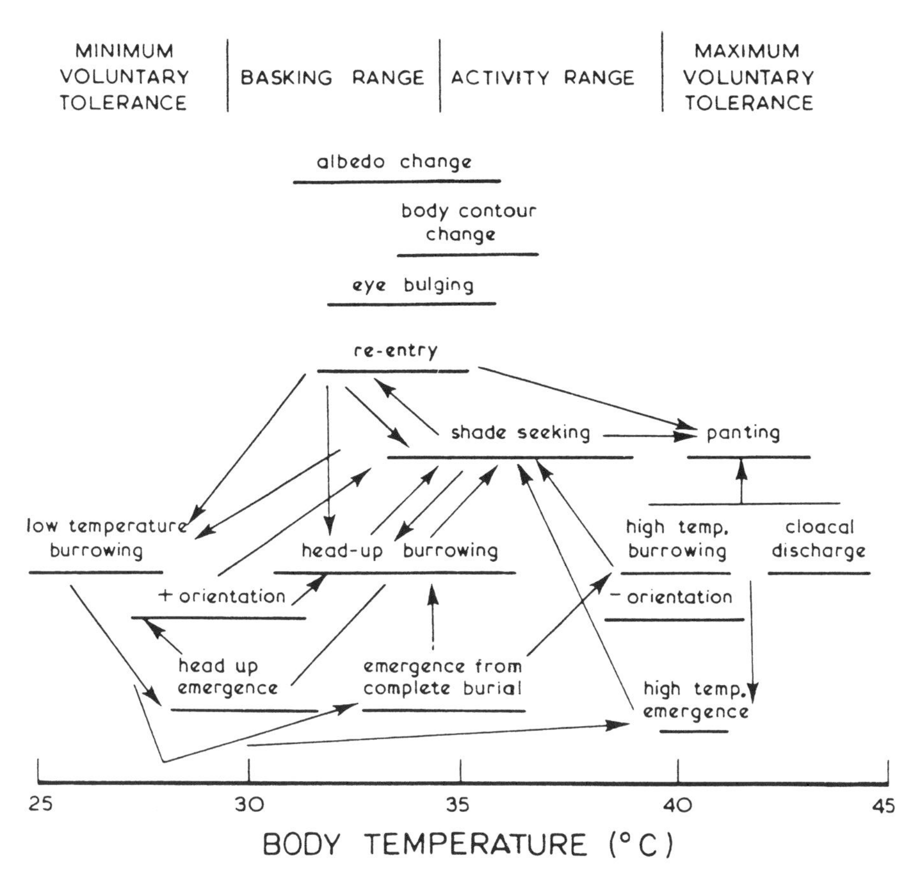


**Fig. 84.** Interrelations of temperature-regulating behaviour in *Phrynosoma coronatum*. The range of body temperatures for each pattern is given. (Cloudsley-Thompson 1971; after Heath 1966)

among Emydidae is thermal control, but a secondary benefit lies in drying the skin and shells (Boyer 1965).

More recently, Hailey and Coulson (1966a) have shown that *Kinixys spekii* is inactive in Zimbabwe during the hottest part of the day, and the length of the midday period of inactivity increases on hotter days. This small species has a mean body temperature (Tb) in the field of 29 °C, which is low compared with that of other tortoises (including the sympatric *Geochelone pardalis*: mean Tb 32.5 °C), but salivates at a similar Tb (38.4 °C). The mean Tb of active *K. spekii* does not vary with sex, type of activity or ambient temperature (Hailey and Coulson 1996b). The physiological control of body temperature in reptiles has been reviewed by Bartholomew (1982), while Avery (1982) provided an essential database for comparative field studies of body tempeatures and thermoregulation. Loveridge (1984) discussed thermoregulation in the Nile crocodile in detail, while Bradshaw (1997) reviewed thermal homeostasis in desert reptiles.

## 10.5 Costs and Benefits of Thermoregulation

The costs and benefits of thermal regulation in lizards have been analysed by Huey and Slatkin (1976). These authors pointed out not only that reptiles thermoregulate by behavioural and physiological adjustments, but also that the resultant control over metabolic processes is generally assumed to be beneficial. Nevertheless, these thermoregulatory adjustments have associated costs which, if extensive, make thermoregulation impracticable. This idea was then extended to an abstract mathematical cost–benefit model which assumed that both costs and benefits are involved in the determination of optimal behaviour. Investigation of the model led to a set of predictions which included the following: (1) the physiologically optimal temperature is not necessarily the ecologically optimal one; (2) thermoregulation is beneficial only when associated costs are low; (3) thermal specialists will normally thermoregulate more precisely than thermal generalists unless costs are high; and (4) lizards will thermoregulate more accurately if the productivity of the habitat is increased or if exploitation competition is reduced. The authors concluded that, where available, data on lizards generally agree with these predictions.

The analysis stressed that thermoregulatory behaviour is extremely complex among lizards. For instance, passivity may, at times, yield greater benefits than thermoregulation; and thermoregulation to particlar, physiologically defined optimal temperatures may even be ecologically maladaptive. The widely held opinion that most lizards thermoregulate precisely (Templeton 1970) may be partially an artefact; this is due to the preponderance of thermoregulatory studies on lizards of deserts or open habitats whose precise thermal regulation is related to the low cost of raising body temperature because basking sites are readily available, and to the necessity of avoiding heat stress at midday during summer. In tropical forests, where many more species of lizards are found, it is probable that precise thermoregulation is less cost-effective and, presumably, much less common.

Pough (1973) outlined the relationship between lizard energetics and diet. Large size requires an herbivorous diet. At weights above about 300 g, lizards tend to be herbivorous, while in the range 50–100 g they are usually carnivorous or insectivorous (Agamidae, Gerrhonotidae, Iguanidae and Scincidae). Heavy Anguidae, Chamaeleonidae, Helodermatidae, Teiidae and Varanidae do not include herbivorous species among their numbers, even species exceeding 300 g in weight. This is due to the presence of morphological, ecological and physiological specializations.

Herbivorous lizards tend to do well in deserts and highly seasonal environments because they are ectotherms, but, because they are also plant eaters, they are, from another point of view, better off in more constant tropical environments. They cannot have their cake and eat it! Modern turtles do somewhat better than lizards in this respect, probably because of their tank-like strategy, which reduces predation, combined with their low metabolic rate (King 1996).

The relationship between temperature physiology and ecology in reptiles has been reviewed by Mayhew (1965), Huey (1982), Peterson et al. (1993) and Garland (1994). Garland (1994) has also discussed the relationship between endurance capacity and body temperature in lizards.

## 10.6 Effects of Cold

Some species of salamanders and anurans can withstand varying amounts of freezing of their body fluids for limited periods of time. In response to the formation of ice in their tissues, they convert glycogen in their livers to glucose which lowers the freezing point of their body fluids. They also produce glucose as a cryoprotectant. The North American wood frog (*Rana sylvatica*), in particular, tolerates freezing of its entire body, which becomes stiff, the eyes turn opaque and both breathing and heartbeat are suspended. Before this occurs, however, the blood glucose level is increased to 60 times above the normal level through the breakdown of glycogen stored in the liver (see Stebbins and Cohen 1995).

The adaptations of reptiles to cold include viviparity, as mentioned in Chapter 8: Hock (1964) pointed out that the five species that range farthest north, *Lacerta vivipara*, *Anguis fragilis*, *Natrix natrix*, *Thamnophis sirtalis* and *Vipera berus*, are all ovoviviparous. Oviparous reptiles could not reproduce in cold regions because the eggs would not develop due to low air temperature, the short summer season or both. Ovoviviparity, however, allows pregnant females to take their internally developing eggs to places where the temperature is higher and to warm them by basking in the sunshine. Even so, the length of the summer season must be about 4 months for successful breeding to take place and some snakes hibernate as soon as they have been born. Although ovoviviparity probably evolved in cold areas, not all cases can be explained by this factor alone because it is seen among urodeles, anurans, aquatic caecilians, as well as some tropical lizards and snakes. The ichthyosaurs, too, were probably ovoviparous. Other factors that may preadapt a group to evolve viviparity appear to be venom in snakes and maternal brooding behaviour (Shine 1995).

Interspecific variation in the critical minimum level depends upon body shape, as does the rate of cooling. The critical minimum temperature can be altered both by seasonal acclimatization and by short-term acclimation, although the physiological explanation of this is not understood (Spellerberg 1973).

Whereas reproductive diversity is more marked among amphibians than reptiles, the latter show a far greater degree of thermal diversity than amphibians. This is due mainly to the fact that the skins of extant amphibians tend to be more permeable than those of modern reptiles. Whether this was the case among Mesozoic forms is open to question. The skin is also an important site of respiration in the amphibians of today. Some of the Plethodontidae are neotenous and reproduce in the larval state and are consequently lungless, while some amphibians that live in cold water have very small lungs.

# 11 Water Balance and Excretion

Water is essential for life. It is a key element for the functioning and evolution of terrestrial organisms, and the most important physiological adaptations of Amphibia and reptiles to life on land are concerned with its use and conservation. The main sources of water loss from animals are through cutaneous transpiration, respiration and excretion. Water is gained by drinking, from metabolism, and from the blood and body fluids of prey. Vegetarian and omnivorous species also obtain moisture from the plant materials they eat. Water balance cannot, however, be considered in isolation because it is intimately bound up with excretion and osmoregulation. All three are concerned with the maintenance of a constant internal environment, or homeostasis, a concept first enunciated by the French physiologist Claude Bernard in his famous aphorism of 1878, 'La fixité du milieu intérieur est la condition de la vie libre, indépendante'. Bernard's contribution to the concept of physiological evolution has recently been evaluated by Bradshaw (1997).

It might be reasonable to suppose that an animal surrounded by an impermeable skin or integument would automatically become independent of water. Such an animal would, however, be incapable of respiration – the uptake of oxygen and elimination of carbon dioxide cannot take place without simultaneous water loss. Cooling the body by the evaporation of water would not be possible either, and it might not be easy to excrete the waste products of metabolism without losing any moisture at all. Homeostasis is therefore dependent upon the maintenance of a balance between the expenditure of moisture in physiological processes and its uptake.

Respiration in reptiles takes place mainly through the lungs, but, in extant Amphibia at least, cutaneous respiration is also important. We shall therefore discuss this matter before consideration of the water lost through transpiration and expended in excretion. Environmental physiologists have long recognized that the exchange of energy, ions, water and respiratory gases is inextricably interlinked, but, for the sake of clarity, these will be discussed as far as possible separately before any attempt is made at a synthesis.

## 11.1 Cutaneous Respiration in Amphibia

As we saw in Chapter 10 the moist skins of extant Amphibia are important in respiration (Duellman and Trueb 1986). The respiratory epithelia of amphibians are located not only on the skin and lungs, but also in the buccal cavity. The relative importance of these sites varies greatly according both to the species concerned and to the environmental conditions. An external gas exchange mechanism must fulfil two functions: to maintain a given oxygen tension in the arterial blood and a given ratio of hydrogen and hydroxyl ions in the body fluids. The ventilation of fishes must be enormous to satisfy the oxygen tension of the blood. Carbon dioxide or carbonic acid concentration is therefore low in fishes so that a proper ratio can be maintained.

On the other hand, the ventilatory requirements of terrestrial animals are not great. Carbon dioxide tension, therefore, tends to increase greatly, requiring immediate adjustment of bicarbonate concentrations or an auxillary mechanism for lowering carbon dioxide tension. Cutaneous respiration may therefore have evolved in Amphibia primarily to cope with rising carbon dioxide concentration, while the more successful land animals eventually adjusted the bicarbonate ion concentration of their blood and developed more efficient lungs (Rahn 1966).

In order to estimate the importance of the skin for eliminating carbon dioxide, the blood pH of the African toad (*Bufo regularis*) was tested after placing experimental animals for varying periods in buffer solutions (pH 7.0 and 4.9) or swathing them in latex rubbers. [pH is a value on the scale 0–14 that gives a measure of acidity (less than 7) or alkalinity (more than 7). A neutral medium has a pH of 7.] The pH of the blood did not vary from pH 7.5 after the toads had been kept in the buffer solutions for 24 h or more. No toads survived for this length of time when swathed in latex, however, but when swathed for periods of 2 h, their blood pH dropped to 6.5. It was therefore concluded that pulmonary and cutaneous respiration are closely adjusted in this species (Cloudsley-Thompson 1970b). This simple experiment does not appear to have been repeated.

The partitioning of gaseous exchange between the lungs and skin and the physiology of amphibian respiration have been studied extensively (Boutilier et al. 1992; Pinder et al. 1992; Shoemaker et al. 1992), and no attempt will be made to review this vast topic in the present book. It is perhaps worthwhile pointing out, however, that in moving from water to land amphibians entered an environment where oxygen is plentiful and moisture potentially scarce. Most species have retained a highly permeable integument and must have ready access to moisture. (Temporary excursions to drier environments will be mentioned briefly below.) A few species with low skin permeability respire almost entirely by their lungs. Amphibian lungs are ventilated by positive pressure from the buccal pump. Breathing air leads to a high concentration of carbon dioxide, but this appears to be compensated for by the presence of bicarbonates which maintain the pH of the body fluids (Shoemaker et al. 1992).

In general, most adult amphibians take up oxygen through both lungs and skin, and excrete most of their waste carbon dioxide through the skin. Since more carbon dioxide is excreted through the skin than oxygen is taken up, decreases in cutaneous gas exchange may affect acid – base balance more than oxygen delivery, as already indicated. Since cutaneous gas exchange is limited to a considerable extent by diffusion, the production of cocoons (see below) decreases cutaneous gas exchange. Not surprisingly, therefore, spadefoot toads do not ventilate their lungs during aestivation. Again, gas exchange in the African bullfrog (*Pyxicephalus adspersus*), whilst in its waterproof cocoon, is almost entirely pulmonary and the blood becomes acidotic (Loveridge and Withers 1981). Cardiovascular physiology in amphibians has been reviewed by Boutilier et al. (1992), Pinder et al. (1992) and Shoemaker et al. (1992).

## 11.2 Cutaneous Transpiration

Water loss through damp skin is naturally very high, especially in the case of small amphibians having relatively large surface to volume ratios. Indeed, these are only to be found in moist tropical regions. The smallest known amphibian is the Cuban arrow-poison frog *Sminthillus limbatus* whose adult size ranges from about 8.5 to 12.4 mm. It lays only one egg at a time. *Psyllophryne didactyla* (Brachycephalidae) from Brazil is a close rival, seldom exceeding a length of 9.8 mm. Several *Eleutherodactylus* spp., and microhylids too, measure only slightly more than 10 mm in length. The smallest toad in the world, *Bufo taitanus beiranus*, originally discovered in the Beira region of Mozambique, does not exceed a length of 24 mm.

A permeable skin creates problems in water as well as on land. Whereas water loss by evaporation across the skin leads to concentration of the body fluids, the uptake of water through osmosis has to be balanced by the excretion of very dilute urine, as described below. Problems of evaporation are especially great in desert amphibians. Indeed, since most amphibians lose water by evaporation much faster than other terrestrial vertebrates, it is not surprising that comparatively few amphibian species have become adapted to life in arid regions.

For example, an experiment carried out on the African toad (*Bufo regularis*) demonstrated this clearly. Toads were placed in desiccators over calcium chloride at different temperatures and weighed at hourly intervals. It was clear from this that the rate of water loss was correlated with the saturation deficiency or evaporating power of the air (the relative humidity was controlled at 10%), and there was absolutely no regulation of transpiration (Fig. 85). Figure 86 shows the relationship between air temperature and mean body temperature in the same species. Over a limited period, evaporation served to control body temperature and a much higher temperature could be

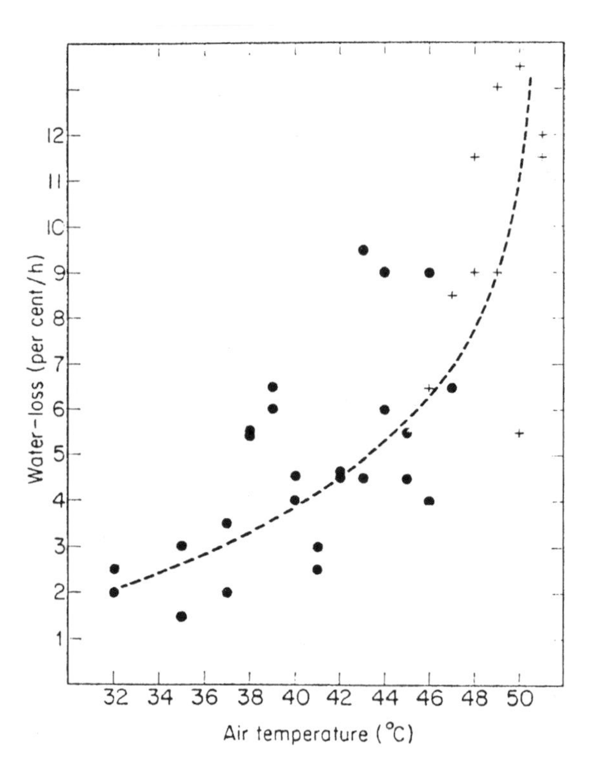


**Fig. 85.** Percentage water loss per hour in dry air from *Bufo regularis* at various temperatures. *Crosses* indicate that the toads died. (Cloudsley-Thompson 1967. Reproduced by permission of the Zoological Society of London)

tolerated in dry air than when the animals were exposed to high temperatures in a container lined with damp filter-paper so that evaporative cooling was prevented (Cloudsley-Thompson 1967). Simple experiments such as this can be carried out by anyone possessing a thermometer and a sensitive balance. Most physiological experiments, however, require the use of more advanced technology.

Heatwole (1984) reviewed the adaptations of amphibians to aridity. He concluded that not only is there no common mode of adaptation, but also it is seldom possible to arrange a series of species into a sequence of increasingly drier habitats which corresponds with an orderly progression in any physiological, behavioural or ecological characteristic. He concluded that evaporation of water takes place through the skin of most amphibians at

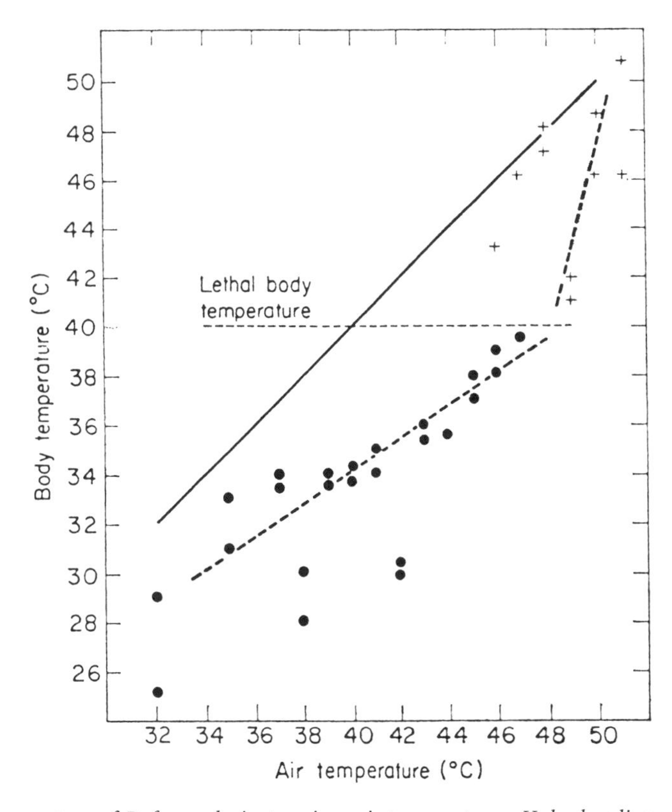


**Fig. 86.** Mean body temperature of *Bufo regularis* at various air temperatures. *Unbroken line* indicates what the body temperatures would have been if there was no evaporative cooling. *Crosses* indicate that the toads died. (Cloudsley-Thompson 1967. Reproduced by permission of the Zoological Society of London)

rates similar to losses from free water surfaces. Only the genera *Phyllomedusa* (Hylidae) and *Chiromantis* (Rhacophoridae) are known to possess mechanisms for resisting evaporative loss from the intact integument.

More recently, Warburg (1997) also concluded that the adaptation of amphibians to the more extreme terrestrial conditions has been achieved in various ways involving modified structures and their functions, behaviour and ecology. He produced a table showing the evaporative water loss (EWL) from a number of desert Anura and Urodela. The lowest rates of EWL were recorded from *Phyllomedusa* spp. and *Chiromantis* spp. – *P. sauvagei* $0.3\,mg\,g^{-1}h^{-1}$, *P. ihrenegi* $0.6\,mg\,g^{-1}h^{-1}$, *C. rufescens* $0.5\,mg^{-1}h^{-1}$ and *C. xerompelina* $0.4\,mg\,g^{-1}h^{-1}$. Warburg (1997) also found that juvenile amphibians usually lose more water and at a higher rate than adults. There is also a tendency for juveniles to aggregate in groups, especially in response to desiccation. Both anurans and urodeles conserve water better in groups than when exposed individually to dry air, and there are apparently seasonal differences in their EWL. Thus, during the wet season, the EWL of the Kenyan reed frog (*Hyperolius viridiflavus*) is 30 times greater than it is in the dry. During the long dry season of the African savanna, large amounts of nitrogenus wastes, chiefly in the form of guanine platelets, are deposited in the iridophores of the skin. These too increase greatly in number, thereby increasing reflectance and reducing heat gain (Kobelt and Linsenmair 1992).

Between the stratum spongiosium of the skin (which contains the dermal glands and muscle fibres) and the stratum compactum (which is composed of collagen fibres) is a layer of mucopolysaccharide deposits known as 'ground substance'. Mucosaccharides are found in all animals to which water retention is important, and are strongly hydrophilous. They can link up with half their volume of water (see Warburg 1997). In some amphibians, the layer of ground substance provides a considerable degree of impermeability, although this is far less than that imparted by the scales of reptiles (Elkan 1968). Surprisingly, this layer gets thicker in older specimens, yet young frogs require even more protection from desiccation (Elkan 1976). This subject obviously merits further investigation.

Many species of anurans can withstand desiccation up to the point at which water loss reaches 50% or even more of their total bodyweight. In fossorial species water can also be stored in and reabsorbed from the bladder. The mechanism involved and the hormonal control of resorption have not yet been investigated. Some fossorial species also form cocoons, as discussed below.

Compared with amphibians, transpiration takes place relatively slowly through the skin of reptiles. In dry air at temperatures in the range 25–40 °C, total water loss from young Nile crocodiles varies from 5 to 20% initial body weight over 24h. The lethal limit of water loss is about 40% (Cloudsley-Thompson 1969). *Alligator mississipiensis* loses 10% body weight in 10h at 38 °C. The Crocodylia is an exclusively amphibious group, so it is not surprising that its members should have integuments that are more permeable to water than those of other reptiles (see below). The situation is quite different in the case of terrestrial reptiles. Although evaporation through the integument constitutes a major source of water loss even from the desert tortoise *Gopherus agassizii*, it is far less than in tortoises from more humid regions.

Cutaneous transpiration is an important factor in the ecology even of desert-inhabiting lizards. It constitutes 39% of the total water loss from *Uta stansburiana* whereas some other species are comparatively impermeable. As Heatwole and Taylor (1987) pointed out, when evaporative losses from reptiles are measured precisely, they are usually found to be of considerable magnitude although lower than in most amphibians. When placed in dry air in a desiccator (above anhydrous calcium chloride), the day-active African skink *Mabuya quinquetaeniata* lost water at an average rate of 5% body weight in 24h at room temperatures of 34 °C (± 1 °C range), while the nocturnal gecko *Tarentola annularis* showed a mean daily loss of only 3.3% under similar conditions (Cloudsley-Thompson 1965). Skinks have a wider range of habitates available to them than geckos. As in tortoises, the impermeability of the integument of lizards and snakes appears to be related to the aridity of the habitat, but there are considerable variations no doubt related to lifestyle, times of activity (Chap. 9) and so on. Evaporation is less at higher humidities: because the saturation deficit of the air used for respiration is not increased by heating, a reptile in its burrow therefore benefits continuously from an advantage in its water economy that is achieved through poikilothermy. Mautz (1982) has provided an extensive review of evaporative water loss patterns in reptiles (see also Mayhew 1968; Cloudsley-Thompson 1971; Lillywhite and Maderson 1982; Minnich 1982; Nagy 1982; Heatwole and Taylor 1987; Bradshaw 1997).

Although the reptilian integument shows low permeability for water, ions and gases, there are numerous exceptions and great interspecific variability (Lillywhite and Maderson 1982). Differences in rates of water loss and survival time have been interpreted as physiological constraints explaining habitat partitioning. Physiological tolerance may adequately predict the ability of a species to colonize habitats with varying climates or to survive periods of drought, but it does not take into account the effects of acclimation or of genetic adaptation (Mautz 1982). Moreover, reptiles regulate their water and electrolyte fluxes both physiologically and behaviourally. Several desert snakes, for instance, make wiping movements with their heads along their bodies. These probably spread lipid secretions which reduce evaporative water loss. Some species of frogs do the same. The Montpellier snake (*Malpolon monspessulanus*) has been shown to possess nasal glands that secrete fluids containing long chain fatty acids. Many reptiles are opportunistic drinkers, selective in their food intake, and hibernate or aestivate when conditions become unfavourable. For further information on cutaneous transpiration the reviews cited above should be consulted.

## 11.3 Cocoon Formation

In 1965, Bill Mayhew noticed that spadefoot toads (*Scaphopus couchi*) captured immediately after emergence from a 2-year dormant period were partially covered with a hard, dry, black material. This appeared to be composed of several layers of skin that had become loosened from the body but not completely cast off. Such a semi-impervious membrane would undoubtedly serve to reduce considerably the evaporation of water from the body. Similar coverings have developed on captive spadefoot toads allowed to hibernate beneath sand in the laboratory (see Mayhew 1968). According to Shoemaker et al. (1992), however, *S. couchi* has not been observed to form an intact, functional cocoon in the sense that the term is now used.

True cocoons have nevertheless been described in other fossorial amphibians by Lee and Mercer (1967) who examined the ultrastructure of the cocoon of the Australian frog *Neobatrachus pictus* (Myobatrachidae). These authors also reported cocoon formation in three species of *Cyclorana* (Hylidae) as well as in *Limnodynastes spenceri* (Myobatrachidae). Functional cocoons comprise many layers of stratum corneum. They have been described in the African *Leptopelis bocagei* (Hyperoliidae) and *Pyxicephalus adspersus* (Ranidae), in the North American *Smilisca baudinii* (Hylidae), as well as in the salamander *Siren intermedia* (Sirenidae). The cocoons of some of these species reduce water loss so effectively that, by absorbing water from the bladder, the animals can survive for several months in dry soil with little change in the composition of their internal solutes (Shoemaker et al. 1992; Stebbins and Cohen 1995).

In his account of the ecophysiology of amphibians inhabiting xeric environments, Warburg (1997) gives a detailed review of cocoon formation in aestivating animals and tabulates the desert species in which it is known to occur.

## 11.4 Ocular Water Loss

The eyes of reptiles that lack 'spectacles' provide free surfaces for the evaporation of water. Spectacles, which have replaced movable eyelids, are present in a number of lizards including the majority of geckos, the limbless pygopodids of Australia, certain skinks, night lizards (*Xantusia* spp.), Lacertidae, and in the burrowing amphisbaenians. They have evidently evolved independently on a number of occasions: therefore although they increase water loss, they must also confer selective benefits, probably by protecting the eyes (Chap. 4).

## 11.5 Respiratory Water Loss

Respiratory water loss in Amphibia is relatively small compared with the amount transpired through the skin. It is, however, almost impossible to separate these two experimentally. In the case of reptiles, on the other hand, respiratory water loss is usually somewhat greater in some species than transpiration through the integument. The ratio of water loss from the lungs to loss through the skin is about 70:30 in rattlesnakes, for example. Mautz (1982) has analysed the patterns of evaporative water loss among reptiles in great detail, and presented extensive comparative tables. Tracy (1982) also gives a table of relative rates of evaporation from the respiratory passages and transcutaneously. The subject is extremely complex because so many factors, both physiological and environmental, are involved. Rates of total water gain and loss from free-ranging animals can be measured using isotopically labelled water, but itemized water budgets can only be obtained from appropriate field and laboratory experiments (Nagy 1982). Respiration and gas exchange among reptiles have been reviewed by Wood and Lenfant (1976).

## 11.6 Excretion

Which came first, the chicken or the egg? Read on and all will be revealed! The function of excretion and osmoregulation is the maintenance of a constant internal environment, irrespective of variations in the environment. The excretory products by which nitrogen is eliminated through the kidneys vary depending upon the amount of water available. Fishes and aquatic amphibians excrete ammonia in the form of ammonium hyroxide, as do the aquatic larvae of terrestrial forms. Ammonia, however, is toxic, and can only be used by animals that live either in water or in environments where water is constantly available.

Like insects and birds, reptiles eliminate nitrogenous waste in the form of insoluble uric acid so that little water is lost in the process. (In a similar way, arachnids excrete guanine which, too, is extremely insoluble.) The evolution of uric acid metabolism is related to the development of a 'cleidoic' or enclosed egg surrounded by a relatively impermeable membrane or shell. Within such an egg, ammonia would soon accumulate and become toxic, but without impermeable membranes the eggs of terrestrial animals would soon dry up and the embryos within would die. Urea, the principal excretory compound of mammals, would not be suitable as the excretory product of an embryo in an enclosed egg because when it became concentrated it would upset the osmotic relations of the developing embryo. (The urea produced by a mammalian embryo passes through the placenta and is excreted by the mother.)

Reptiles, like insects and birds, are therefore uricotelic and have evolved the necessary physiological 'machinery' to excrete uric acid in the egg stage: they retain this useful ability throughout their lives. So, the answer to the question as to which came first, the chicken or the egg, must be the cleidoic egg! Adult amphibians that live on land, like mammals, tend to excrete urea. For this reason they are usually dependent upon a relatively moist environments. (Some of the adaptations of amphibians to arid environments have already been mentioned. Others will be discussed later.)

It must be stressed that the above account is greatly simplified. Aquatic amphibians have a low level of plasma urea and a high ammonia content in their urine (Schmid 1968). Most of the ammonia is formed in the kidneys, but some may be excreted through the skin. The amount varies from 90% in *Necturus maculosus* to 15% in *Xenopus laevis*. Adult newts excrete a higher proportion of ammonia when in water during the breeding season than they do on land, but most terrestrial amphibians, including fossorial species and those that inhabit xeric environments, as we have seen, are uricotelic (Warburg 1997).

Most terrestrial reptiles are also uricotelic, but aquatic crocodilians and marine turtles mainly excrete ammonia and urea. There appears to be a gradation among chelonians from amphibious species to xerophilous ones which excrete mainly uric acid. Lizards and snakes are not strictly uricotelic, however, and the proportions of uric acid and urea may vary considerably. Land tortoises show even greater temporal variations in the ratio of urea to uric acid secreted, and changes in the ratio have been observed within a single individual of *Testudo leithii*, a North African species, and of the sub-Saharan *Geochelone sulcata*. Excretion in reptiles has already been reviewed on a number of occasions (Cloudsley-Thompson 1971; Dantzler 1976; Bradshaw 1997).

## 11.7 Water Uptake

Anurans are unique among vertebrates in that cutaneous absorption is their most important means of water uptake. Many desert species possess a conspicuous 'seat patch' which is a region of thin skin on the ventral surface through which water can be absorbed from the substrate. This takes place very rapidly after dehydration. The process is not directly affected by temperature although there is a relationship between the degree of water loss and the rate of uptake (see discussion in Warburg 1972, 1997; Duellman and Trueb 1986; Boutilier et al. 1992; Stebbins and Cohen 1995).

Most amphibians and reptiles drink free water when this is available, but some reptiles obtain essentially all their water from their food. Water is also produced as a by-product of metabolism, but it is unlikely that this is significant because, as the metabolic rate increases, so does respiration and consequent respiratory water loss. The two tend to cancel each other out. A potential source of water is condensation on the exterior surface of the body as well as in the respiratory system of an animal whose body temperature is below the dew point of its environment. Whether or not this actually occurs has not been ascertained, but the sidewinding viper *Bitis peringueyi* of the Namib desert dunes responds to moisture from the advective sea fog engendered by the cold Benguella current by flattening its body against the substrate. This increases the surface area on which moisture condenses, and the droplets are licked up by the snake (Louw 1972).

At one time it was believed that *Moloch horridus, Uromastyx hardwickii, Cordylurus giganteus* and other desert lizards possessed hygroscopic skin that absorbed water like blotting paper. It is now known, however, that although the skin of *M. horridus* is indeed like blotting paper, it possesses a number of fine capillaries which lead to the mouth. The lizard appears to secrete a hygroscopic mucus from glands at the edge of the lips. This absorbs moisture from dew and scattered showers, and is then swallowed. The subject has been reviewed by Cloudsley-Thompson (1971), Minnich (1982) and many others.

It has long been thought likely that crocodiles might be able to absorb water through the skin and cloaca (Cott 1961; Diefenbach 1973). In the case of the Nile crocodile, however, this has been disproved. Young animals were partially desiccated and their bodies then submerged in water, except for their heads, and they were prevented from drinking. After their bodies had been dried, it was found that there had been no gain in body weight (Cloudsley-Thompson 1969). These observations contrasted with results obtained by Bentley and Schmidt-Nielsen (1965) who claimed that 70% of the water taken up by *Caiman sclerops* was absorbed through the integument and that differences between the two sets of results might have been due to the fact that the crocodiles were partially dehydrated whereas the caimens were fully hydrated and living in water. They were only losing water in their urine and by respiratory evaporation, and their total uptake was low. At the same time, Bentley and Schmidt-Nielsen (1965) used a different technique which involved dissolving phenosulfurphthalin (phenol red indicator) in the bathing solution. After this had been recovered and made alkaline, they measured calorimetrically the amount that had accumulated in the alimentary canal. Some years later, Diefenbach (1973) confirmed gravimetrically that neither the Nile crocodile nor *Caiman crocodilus* – at temperatures between 26 and 31 °C – gain water hygroscopically through the skin but only by drinking. He pointed out that phenol red is absorbed by the gut and later excreted. Consequently, its presence in the gastrointestinal tract does not provide a quantitative indication of the amount of water drunk. Nevertheless, it is now generally recognized that the permeability of the reptilian integument increases considerably more when the skin is in contact with water than when it is exposed to air.

Bidirectional water flux through the skin has subsequently been confirmed using tritiated water molecules, both in whole animals and in vitro skin preparations of several species of freshwater turtles and snakes. It is clear that the integument is the major route of both water influx and efflux in these aquatic reptiles. In the case of the dog-faced water snake (*Cerberus rynchops*) of Southeast Asia, which sidewinds on tropical mudflats and catches fishes underwater, fluxes across isolated skin account for all body water exchange. Seawater contains some 3.5% salt which is about three times as much as in the blood and body fluids of vertebrates. There is, therefore, a potentially serious osmotic problem for marine reptiles. Osmosis is the movement of water from a region of low solute concentration to one of higher concentration through a semi permeable membrane – that is, a membrane whose structure allows the passage of a solvent (water) but prevents that of a solute (salt).

Elapids solve this problem by retaining freshwater, and at the same time excluding saltwater and excess salt through nasal salt-excreting glands (see below; for references see Lilleywhite and Maderson 1982; Heatwole 1987; Heatwole and Taylor 1987; Greene 1997).

On land, many reptiles including the semi-aquatic Nile monitor (*Varanus niloticus*) and worm lizards (*Rhineura floridana* and *Anniella pulchra*) have been shown experimentally to be unable to absorb moisture through the skin from a humid substrate. This is a very different matter, however, from absorbing water when the integument is fully hydrated and immersed in it! Although there appear to be many exceptions, the reptilian integument, in general, shows low permeability for water, ions and gases. There is much scope, however, for further research on the topic.

## 11.8 Osmotic Balance

Maintenance of a constant internal environment, and the osmolarity and ion concentration of the body fluids, has been discussed by many authors (for references see Shoemaker et al. 1992; Warburg 1997). In general, aquatic amphibians have a less concentrated plasma and therefore lower osmoregularity than terrestrial species. When the animals are dehydrated, however, both the osmolarity and ionic concentration increase markedly. Nevertheless, somewhat surprisingly, the amphibians of arid and semi-arid environments do not appear to differ significantly in their water balance from those that inhabit mesic regions, although they can apparently survive for longer without water and absorb it more quickly and in larger quantities when it becomes available.

Aquatic amphibians have a low level of urea and more ammonium hydroxide in their urine than species from more arid environments. Spadefoot toads accumulate high concentrations of urea during dormancy, while some arboreal frogs of xeric regions are able to produce either urates or uric acid, as we have seen. The endocrinological control of water and nitrogen balance in Amphibia have been the subject of intensive research, and major reviews are listed by Shoemaker et al. (1992) and Warburg (1997).

As in amphibians, the state of hydration of reptiles has an important influence on the osmotic concentrations of their plasma, specific solutes especially sodium chloride, and a number of other internal parameters. Reptiles tolerate the accumulation of sodium salts in their body fluids more easily than they tolerate those of potassium. Freshwater species can regulate their plasma solutes while they are feeding and exposed to adequately warm temperatures. Even marine reptiles can maintain sodium plasma concentrations for long periods in freshwater. Marine Chelonia have relatively high plasma osmotic concentrations compared with freshwater and terrestrial species. The increased osmotic concentrations that accompany the acclimation of freshwater terrapins to seawater result mainly from water loss rather than from gains in sodium chloride. They do not appear to be a response that creates a more favourable osmotic gradient between the terrapins and the seawater (see Minnich 1982; Heatwole and Taylor 1987).

Crocodylids, through a suite of morphological specializations, are better adapted than alligatorids for life in hyperosmotic, saline environments. The presence of such specializations, even in freshwater crocodiles, is evidence of a marine phase in their evolution. Crocodiles are better able than alligators to discriminate freshwater from hypertonic (more saline) seawater and to avoid drinking the latter. This is an important osmoregulatory behavioural mechanism for estuarine crocodiles (Jackson et al. 1996).

The survival of terrestrial reptiles, and especially of desert reptiles in periods of drought, depends upon the regulation of potassium ions as well as on the ability to excrete excess electrolytes with little water loss; the nasal lachrymal salt gland is capable of excreting a very concentrated brine. Reptilian salt-excreting glands have evolved independently in several groups of marine, estuarine (Fig. 87) and terrestrial reptiles. In sea snakes they are sublingual, in terrestrial lizards nasal. Their presence has not been confirmed in *Sphenodon* spp. or in Crocodylia. Salt-excreting glands have been reviewed by Dunson (1976) and by Minnich (1982), while osmoregulation in reptiles is the subject of a review by Bentley (1976).

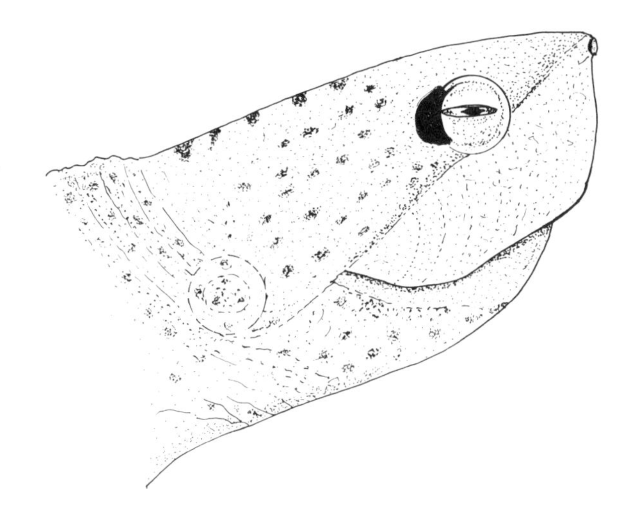

**Fig. 87.** Head of the estuarine diamondback terrapin (*Malaclemys terrapin*) with the post-orbital lachrymal salt gland shown in *black*. (After Dunson 1976)

## 11.9 Evolutionary Recolonization of the Water

When the first reptiles evolved, they were apparently well adapted for terrestrial life (Carroll 1969a,b). Nevertheless, many types have since reinvaded the freshwater habitats of their amphibian ancestors and others have recolonized the seas. The latter include icthyosaurs, nothosaurs, plesiosaurs, mosasaurs, turtles, crocodilians and so on, whose morphological adaptations for life in water have already been outlined (Chap. 5). Reinvasion of aquatic environments is not surprising because the design of reptiles incorporates many physiological properties that preadapt them to life in water. These have been reviewed by Seymour (1982), and include ectothermy, a pronounced capacity for anaerobic metabolism, tolerance of severe acid–base disturbances, ventilation involving holding the breath, and a circulatory system characterized by a degree of intraventricular shunting and vascular responses that are invoked in the diving response. Some totally aquatic reptiles exhibit vital physiological functions that are scarcely different from those of their terrestrial relatives. (This subject has been reviewed by Seymour 1982.)

This and the preceding chapter have been devoted to physiological topics which have attracted an astounding amount of original research in recent years because reptiles are an ideal taxon for studying ecological physiology. In this book, I have merely outlined some of the main ecophysiological adaptations of amphibians and reptiles that have enabled them to exploit a wide range of habitats, and no more than highlighted some of the major reviews that have been published on various aspects of their environmental adaptations. The reason is that modern physiological research often requires the use of highly specialized techniques and apparatus which are beyond the reach of the average amateur naturalist.

# 12 Relationships with Mankind

Amphibians and reptiles play an important role in the energy flow and nutrient cycling of many ecosystems, including even those of arid lands. In numerous habitats they convey energy from the small invertebrates on which they feed to the larger vertebrates that prey upon themselves. As ectotherms, they require low rates of energy for their own maintenance and, unlike endotherms, they lose little metabolic energy in the form of heat. Consequently, they act as a reserve of energy within the various ecosystems in which they occur.

It is not, however, with their indirect ecological importance to mankind or their role in the control of insect pests that this chapter is concerned, but, rather, with their direct relationships. As I hope this book has already made clear, amphibians and reptiles are surprisingly interesting and often astonishingly beautiful. Most people who study them do so for intellectual rather than economic or medical reasons. Nevertheless, the latter should not be discounted and will be discussed below.

## 12.1 Sources of Food and Other Products

The legs of frogs, an important source of protein in some impoverished societies, are also regarded as a delicacy by gourmets in various parts of the world. Affluent countries import amphibian products in large amounts. France imports some 3.4 million tonnes of frogs' legs per year, mostly from Indonesia and Bangladesh, while the United States imports between 1000 and 2000 tonnes annually. Bangladesh has joined India and China in controlling the harvest because of a consequent increase in the numbers of mosquitoes and other noxious insects (Stebbins and Cohen 1995).

Little is known of the extent to which defensive chemicals are transmitted from one animal to another in a food chain, and even frogs' legs may be implicated. After eating the meat of frogs captured locally, French soldiers in Algeria, during the last century, sometimes developed urogenital symptoms including 'érections douloureuses et prolongées', such as were known to doctors to be characteristic of cantharidin poisoning. At one time, before its dangerous properties had been recognized, cantharidin prepared from the wing cases of 'Spanish fly' or blister-beetles (*Lytta* spp.) was used as an aphrodisiac. Apparently the Algerian frogs had been collected from areas where blister-beetles were abundant, and examination of their gut contents showed that they had been feeding on the beetles. The legionnaires were therefore getting more than they had bargained for – a potent aphrodisiac along with their epicurean delicacies (Cloudsley-Thompson 1980b)!

Reptiles are, or have been, a much more important source of human food than amphibians. Green turtles, in the form of turtle soup, until recently, formed an important course in many a British Royal banquet or Livery dinner. As a member of the Livery of the Worshipful Company of Skinners in the City of London, I can remember how delicious it was. Thank goodness marine turtles are now protected species! Fortunately, too, modern plastics have rendered tortoiseshell almost redundant. According to Pope (1955), until they were protected, the most favoured turtles in the United States were diamondback terrapins (*Malaclemys terrapin*), followed by snapping turtles (*Chelydra serpentina*). Soft-shelled turtles were relished throughout the world and especially in Japan, China and the United States, while turtle eggs were and in some places are still gathered in vast numbers for human consumption.

After the use of turtle meat and eggs for food came the use of their shells for furniture inlay and other decorative crafts. The hawksbill turtle (Fig. 42) was the main provider of commercial tortoiseshell. Only the top part of the shell, however, shows the richness of markings for which it has always been prized. When its owner is alive, however, the turtle's shell is drab in appearance, and often discoloured by the presence of algae and barnacles. Hawksbill turtles were highly pized in ancient Egypt and their shells used for the manufacture of bracelets and knife handles. Illegal trade in turtle products unfortunately continues to take place despite international attempts to control it (Alderton 1988).

The flesh of the tail of the young alligator was described by Christian Schultz in 1808 as being delicious, and many other travellers in America during the nineteenth century thought the same (Neill 1971). Although crocodiles too are said to be quite edible, their meat is not used in the Western world. Crocodilian hides, however, are a source of excellent leather used extensively for the manufacture of shoes, travel bags, belts and so on. Considerable waste is invoked, however, because the bony plates on the dorsal surface render the hide from that region quite valueless. Even the belly skin of some species is useless for the same reason. The introduction of crocodile farming has gone far to preserve many endangered species without affecting their use in trade.

Tanned lizard and snake skins, unfortunately, also produce high-quality leather used for briefcases, shoes and the like. Lizards and snakes are also eaten in many parts of the world without having been reared in captivity. The chuckwalla is relished by some Amerindians, and spiny-tailed lizards (*Uromastyx* spp.) are enjoyed by the bedouins of the Arabian peninsula (Pope 1955). Snake meat is regarded as a delicacy in Southeast Asia where the Indian python (*Python molurus*) is especially esteemed by the Chinese, while to the Papuans, as well as the Australian aborigines and some South American tribes, snake meat is an important source of protein. Rattlesnakes of various species are regarded as a delicacy in the United States and Central America where they are sometimes killed in unpleasant and cruel rituals (Engelmann and Obst 1984).

## 12.2 Use in Medicine and Research

Both amphibians and reptiles have been studied extensively by herpetologists and have contributed to numerous fields of investigation in ecology, space research, telemetry, cybernetics, social organization and ageing. Indeed, they have proved themselves to be excellent material for research on various physiological problems, as already explained in the *Introduction*. Neill's (1974) *Reptiles and Amphibians in the Service of Man* outlines some of the innumerable biological questions for which they have supplied satisfactory explanations or shed illumination. These include evidence for continental drift or plate tectonics and the physiology of sleep. I shall not, therefore, discuss them further here.

Basic research, undertaken solely for the acquisition of knowledge, has often proved more successful in innovating new developments beneficial to human welfare than has research directed specifically towards anthropocentric objectives. Studies of behaviourally controlled fever in amphibians and reptiles are a case in point. When injected with pathogenic bacteria, ectotherms voluntarily raise their body temperatures by moving into warm locations. If prevented from doing so, more of them succumb to infection than when they are allowed to raise their body temperatures. This finding has contributed to a re-evaluation of the traditional methods of treating fevers in human beings.

Amphibian skin secretions are often strongly antibiotic because they have to protect the moist integument from infection by bacteria, fungi and other pathogenic microorganisms. Skin secretions that afford protection from predators are used for arrow poison by the Amerindians of South America. In some cases, they have curare-like painkilling properties many times more effective than morphine, and research is currently being directed towards analysing and synthesizing the chemicals responsible. Some secretions have psychedelic effects that may prove to be of value in the treatment of mental disturbances; others may eventually be used to counteract Alzheimer's disease and depression (Stebbins and Cohen 1995).

Snake venoms have likewise contributed to important developments in physiology and medicine. They are also of considerable importance on account of the public health problem of snakebite. Various aspects of these topics are discussed in Thorpe et al. (1997). The venoms of cobras, pit-vipers, coral snakes and the Gila monster (*Heloderma suspectum*) (Fig. 49) have all been and are still being used in homeopathic medicine, while many hundeds of articles on the employment of venoms in regular medicine have been published during the present century. These include their use in the treatment of haemorrhagic disorders, severe nose bleeds, uterine haemorrhage and so on, epilepsy, leprosy, and in palliating the symptoms of pain in people suffering from inoperable cancer, angina pectoris and migraine (Minton and Minton 1971).

## 12.3 Dangerous Reptiles

Amphibians are only dangerous to mankind when their skin excretions are used as arrow poisons, and no Chelonians are harmful. Since 1943, however, there have been 225 attacks by alligators on human beings in Florida, seven of them fatal. The crocodilian that is probably most dangerous to mankind is the Indopacific sea-going *Crocodylus porosus*. Only very large individuals normally become man-killers and, in general, crocodiles do not usually attack human beings. Pooley et al. (1989), however, point out that circumstances occasionally occur which permit the animals to display grossly atypical behaviour. Towards the end of World War II about 1000 Japanese soldiers, trying to escape from Ramree Island through a mangrove swamp to the mainland of Burma 30km away, were attacked at night by estuarine crocodiles and only about 20 survived. The event was described by an American biologist, Bruce Wright, who was a member of the British forces that had trapped them; 'scattered rifle shots in the pitch black swamps punctured by the screams of wounded men crushed in the jaws of huge reptiles, and the blurred worrying sound of spinning crocodiles. . . . At dawn the vultures arrived to clean up what the crocodiles had left'. The story has, however, been queried.

South-East Asia and northern Australia are home to a number of crocodilians of impressive size. Yet I have heard of bathers near Darwin actually leaving their towels on the notices warning them that the waters were inhabited by crocodiles! The motivation behind crocodile attacks may not be simple and in some instances is not a feeding response. In a recent study of 27 cases of fatal crocodile attacks in Australia, there was evidence that all or part of the victim's body had been eaten on 16 occasions, in eight the body was never recovered and possibly eaten, while in three the body was recovered intact.

Both Australian and Nile crocodiles show aggression towards boats that enter their territory, and attacks by them may be categorized as defence of territory, hatchlings or nests, self defence and predatory. Like sharks, crocodiles are rarely seen before they attack; they are not deterred by noise and there is no safety in numbers as far as human beings are concerned. American alligators are largely fish eaters and there is little evidence that they attack animals as large as human beings on a regular basis. Again, defence of territory, mates or nests appear to be frequent causes of incidents involving human victims.

The Indian mugger (*Crocodylus palustris*) has seldom been known to kill a human being although the Hindu custom of burning the dead on the banks of the Ganges and other rivers and afterwards throwing the remains into the water gives crocodiles many opportunities to feed on human flesh. There is no record of an Indian gharial or Johnston's crocodile (*Crocodylus johnsoni*) of northern Australia having killed a human being, but black caimans (*Melanosuchus niger*) have been known to defend their nests (Guggisberg 1972). Even estuarine and Nile crocodiles make excellent pets when babies, and the dangers posed by crocodilians in general are often much exaggerated. Nevertheless, it is unwise to take any unnecessary risks with these potentially dangerous animals.

Although the venom of *Heloderma* spp. is painful and potentially dangerous, the beaded lizard and gila monster (Fig. 49) seldom bite human beings. Indeed, the only dangerously venomous reptiles are cobras and their allies (Elapidae), vipers (Viperidae) and some back-fanged Colubridae such as the African boomslang (*Dispholidus typus*) and twig snake (*Thelotornis kirtlandi*) (Fig. 29), the mangrove snake (*Boiga dendrophila*) and long-nosed tree snake (*Dryophis nasuta*), both from Southeast Asia, and the lyre snakes (*Trimorphodon* spp.) of North America.

Bites by vipers are followed by severe pain and swelling in the region of the injury. Blood-stained serum may ooze from the fang marks and enter the subcutaneous tissues causing discoloration of the skin. Clotting of the blood is inhibited so that there may be haemorrhage of the lungs or intestine, and the patient coughs up blood or passes it through the rectum. Small purple spots often appear beneath the skin where the blood has leaked from damaged vessels. Later, areas of tissue become gangrenous and are sloughed away. When death occurs, it is usually due to failure of the heart or respiration. Although cobra venom not infrequently also causes severe local damage, there is little pain at the site of the injury. The venoms of most snakes contain a variety of both neurotoxins and blood poisons, but the former tend to predominate in Elapidae, the latter in Viperidae. Moreover, neurotoxins may also act on the blood system, while blood poisons can have side effects on the nervous system, so that the effects of both are complicated, Nevertheless, muscular weakness, as well as depression of the breathing and of the heart, are more characteristic of the effects of elapid venom, and death usually follows much sooner than in the case of viperid poisoning.

Surprisingly, living cobras are being used in Indonesia for crowd control. A police officer wielding a cobra may lower the spirits of rioters and persuade them to back off! Cobras do not normally strike unless disturbed. Nevertheless, anyone who works on or indeed keeps venomous snakes in captivity should take the greatest care at all times and *never* handle dangerous charges more than is absolutely necessary.

## 12.4 Mythology and Folklore

When, in *As You Like It*, Shakespeare wrote 'Sweet are the uses of adversity, which, like the toad, ugly and venomous, Wears yet a precious jewel in his head', he was repeating a common myth of the time. John Bunyan, too, was among those who in days of old gave this remarkable fable their support. In his apology for *The Pilgrim's Progress*, he stated: 'If that a pearl may in a toad's head dwell, And may be found too in an oyster-shell, If things that promise nothing do contain What better than gold – who will disdain That have an inkling of it there to look That they may find it.'

It used to be a common idea also that toads were venomous. Edmund Spenser, for instance, wrote: 'The grieslie tode stoole growne mought I se, And *loathed* Paddocks (toads) lording on the same,' and Alexander Pope, in one of his satires, mentioned the toad's alleged habit of spitting venom: 'At the ear of Eve, familiar toad, Half froth, half venom spits himself aboad'. Again, Thomas Chatterton, 'Ye toads, your venom in my footpath spread' and even Robert Browning 'Creatures that do people harm – The mole and toad and newt and viper' regarded the toad with disfavour. As late as 1768, Gilbert White wrote in *The Natural History of Selborne* that 'the matter with regard to the venom of toads has not yet been settled'.

In the past, credence was given to the idea that frogs and toads were able to live for ages in solid blocks of stone without the benefit of food reaching them. Another superstition reflects the belief that during spring, myriads of small frogs sometimes fall from the clouds. The ridiculous idea that salamanders were able to live in fire was at one time commonly accepted as a fact of nature, and was handed down from Aristotle and Pliny! In comparatively recent times, newts or efts were blamed by country folk who suffered from indigestion and assumed that their abdominal pains were being caused by a newt which they must inadvertently have swallowed in the egg stage. Illness in cattle was sometimes attributed to the same cause and extraordinary means were adopted to get rid of the hypothetical pest. Frogs still play an important role in South American Amerindian mythology, as do tortoises and squamates. Petroglyphs of tortoises with a human foot in the centre of the shell are common, and reptilian forked hemipenes are painted on houses.

Not surprisingly, the mythology of reptiles, especially snakes, is considerably more extensive than that of amphibians. Chelonians have featured in various stories which reflect their slowness and determination. The origins of Aesop's well-known fable of the tortoise and the hare appear to reside in various legends found throughout the continent of Africa. Among the Akamba people it is a race between a tortoise and a fish eagle to win the hand of a Kamba girl; among the Hottentots, the losers are ostriches. Chelonians also feature predominantly in creationist tales. The Buriats of central Asia relate how, in the beginning, there was no land, only water. God then decided to turn a large turtle onto its back, thus creating land. In another story a huge turtle supported the land on its shell. In some parts of India, turtles are still revered today as sacred creatures (Alderton 1988).

The metaphor of crocodile tears has long been used to indicate hypocritical lamentation. In the Middle Ages, travellers returning from foreign countries described how crocodiles wishing to lure their human prey to destruction would beguile them with tears and then devour them with the same manifestations of regret. Francis Bacon expressed this in the words, 'It is the wisdom of crocodiles that shed tears when they would devour', and Shakespeare made Queen Margaret say: 'Henry, my lord, is cold in great affairs, Too full of foolish pity, and Gloucester's show Beguiles him as the mournful crocodile With sorrow snares relenting passengers'. John Dryden used the same simile in his play *All for Love* – 'Caesar will weep, the crocodile will weep, To see his rival of the universe Lie still and peaceful there' – when Antony described Octavia bearing his cold ashes in her widowed hand. Tennyson referred to crocodiles weeping in his poem *The Dirge* and Sir John Suckling wrote of a certain lady, 'Hast thou marked the crocodiles weeping Or the foxes sleeping? . . . Oh! so false, so false is she!'. Coming to more recent times, as a small child the crocodile was my favourite animal and I used to speak of it as the 'tick-tock', a reference to the crocodile in J. M. Barrie's *Peter Pan* that had swallowed an alarm clock!

Crocodiles were venerated in Egypt from earliest time. Strabo (63 B.C.–A.D. 20), impressed by his great predecessor Herodotus, wrote much of the worship of Sebek, the water god whose living incarnation was the Nile crocodile and who is therefore always represented with the body of a man and the head of a crocodile. Mummified crocodiles have been found in many rock tombs, those of Fayum being of particular importance to Egyptology because they were packed in papyri. Near Thebes many had scales perforated where golden rings had been fastened.

On the other hand, Herodotus emphasized that some Egyptians regarded crocodiles with considerable hostility, believing that, after the murder of Osiris, Sebek had helped Set to escape by allowing him to shelter within the body of a crocodile. As late as A.D. 335, sacred crocodiles were still being fed by priests at Arsinoë, the ancient Crocodilopolis. The crocodile cult probably originated far back in pre-dynastic times and ramifications have appeared among the Ashanti of Ghana, on the Ivory Coast, Nigeria, Liberia, Zaire and East Africa. Guggisberg (1972) discusses the subject in some detail.

Many African myths about crocodiles have subsequently been applied to the American alligator according to Neill (1971), who devoted a chapter of his book *The Last of the Ruling Reptiles* to crocodilian legends. Some of these so-called legends are, in fact, no more than the truth. In sifting truth from chaff, Neill cast doubt upon some aspects of crocodile behaviour that later have been reliably observed and reported; for example, that young Nile crocodiles call their parents at hatching time or that crocodilians indulge in territorial battles. Nevertheless, his is a fascinating book warmly to be recommended to herpetologists. Garry Trompf (1989) also devoted a chapter to mythology, religion, art and literature. He concluded that the creature referred to as 'Leviathan' in the Book of Job was the crocodile, while the dragon of ancient China may have been the Chinese alligator (*Alligator sinensis*). In Southeast Asia, the crocodile used to be regarded as the reincarnation of a departed ruler continuing to impose his fierce authority and punitive powers. In West Timor the princes of Kupang were reported, as late as 1884, to sacrifice perfumed and prettily dressed young girls to the crocodiles who were believed to be their ancestors, while, on Borneo, the Kayan thought that the crocodile was actually a guardian angel who could become one's blood relation! Crocodiles are prominently featured in Melanesian sculpture and in the Aboriginal wall paintings of tropical Australia.

Lizards play a smaller role in mythology than do crocodiles or snakes; but the supposition that chameleons live on air is of great antiquity and dates back at least as far as the days of Moses. The chaemeleon was listed among the unclean creeping things mentioned in the book of Leviticus. In the Authorized Version of the Bible it is translated from a word meaning 'to breathe', which, in the Revised Version, is rendered as 'chameleon'. Hamlet was referring to the same idea when he answered the King's enquiry about his health with the words, 'Excellent, i' faith, of the chameleon's dish; I eat the air promise crammed: you cannot feed capons so'. Considerably later, Shelley was to write: 'Cameleons feed on light and air; Poets' food is love and fame!'. He also referred to the chameleon's faculty to change its colour.

Another lizard, the misnamed slow-worm (*Anguis fragilis*), was thought to be venomous. Shakespeare called it 'the eyeless venom'd worm' and mentioned one of the ingredients of the witches' cauldron as being 'a blindworm's sting'. Like snakes, with which they were confused, slow-worms were invariably thought to be slimy (see below).

The spread of Christianity throughout the ancient world marked the beginning of an especially difficult period for snakes. They had always been disliked by human beings because some species are venomous and all were 'tarred by the same brush'. However, when the serpent made its appearance in the book of Genesis, beguiling Eve to violate the commandment of God, it created a very bad impression on future generations of humans – although Eve herself was apparently in no way put out when the serpent sidled up to her in the Garden of Eden to make enticing comments about the Tree of Knowledge! Many artists have since portrayed the scene, depicting the snake realistically; others have portrayed the animal in the form of the devil but with the body of a serpent: surprisingly, the reptile had lost its legs before it was cursed! So a myth was created that justified the persecution of snakes, and there is probably no other animal in the world about which so much nonsense has been spoken and written.

In the presence of the Pharaoh of Egypt, Aaron's rod was transferred into a snake. It is true that snakes can be hypnotized so that they are unable to move, but, even if this were the true explanation of the miracle, it would not explain Aaron's snake swallowing up other snakes conjured up by sorcerers in the service of Pharaoh – unless, of course, Aaron's serpent was not only extraordinarily phlegmatic but also an ophiophagous species!

A 'serpent of brass' was set up by Moses to rescue the people from 'fiery serpents', as described in the book of Numbers. However, the latter were, in fact, probably guinea-worms (*Dracunculus mediensis*: Nematoda), common parasites of tropical Asia and Africa which live in the deeper layers of the skin and are in no way related to snakes. They are acquired through inadvertently swallowing small crustaceans (*Cyclops* spp.), the secondary host, in unfiltered water. Although not known from Sinai, the Israelites might home suffered from them elsewhere. In his Revelations, St John described the seizing and binding of 'that old serpent which is the Devil' and Christ condemned the Pharisees with the words 'Ye serpents, ye generation of vipers'. Other Christian legends are related by Engelmann and Obst (1984).

Although most people know that Cleopatra died from the bite of an 'asp', there has not been agreement as to which species of snake that was. The name asp is today applied both to the Egyptian cobra (*Naja haje*) and to the Saharan horned viper (*Cerastes cerastes*). The snake which Cleopatra applied to her bosom after having 'pursued conclusions infinite of easy ways to die' could, theoretically, have been either of these species, or perhaps the common sand viper (*C. vipera*) of North Africa.

Cleopatra VII was co-ruler of Egypt with her brother Ptolemy XIII from 51 B.C. until he ousted her in 48 B.C. Restored the following year by Julius Caesar, she accompanied him to Rome, returning to Egypt after his assassination in 44 B.C. Here she met Mark Antony, who abandoned his wife Octavia in order to live with her. The two were defeated by Octavian, who later became the Emperor Augustus, at the battle of Actium in 31 B.C. and both committed suicide in 30 B.C. Shakespeare's reconstruction of their story follows the romantic account in Plutarch's *Parallel Lives*.

During the late seventeenth century a lively controversy revolved around the nature of the viper's bite. According to the famous Italian biologist Francisco Redi, its dangerous effects were caused by the yellowish fluid that flowed from the fangs. On the other hand, M. Charas, a French chemist, ascribed it to the snake's 'enraged spirits'. The latter view was the more popular and, in the final scene of *Antony and Cleopatra*, Cleopatra says to the asp: 'With thy sharp teeth this knot intrinsicate of life at once untie. Poor venomous fool, Be angry, and dispatch . . .'. Redi's opinion was based on scientific experiment and eventually prevailed. Charas was not completely wrong, however, for an irate snake usually injects more venom than does one whose anger has not been aroused. About the same time Edmund Spenser, the English poet, wrote of '. . . the stings of aspes that kill with smart'. The venom of a viper is very much more painful than that of a cobra, which is why criminals who were executed by the bite of a cobra were regarded as receiving a 'favour', and Cleopatra is therefore more likely to have chosen a cobra than a viper as the instrument of her death. Moreover, the cobra was a symbol of royalty in ancient times and, during the Greco-Roman period, was used for the execution of criminals. This is further evidence in support of the hypothesis that Cleopatra's asp was a cobra. Finally, the Egyptian cobra is much larger and produces more venom than the horned viper. Consequently, its bite is the more likely to be fatal – an important consideration to anyone determined to die.

After the death of the Queen, one of Caesar's guards in *Antony and Cleopatra*, who made an examination of her apartments, exclaimed: 'This is an aspic's trail, and these fig leaves Have slime upon them such as the aspic leaves Upon the caves of Nile'. In this, Shakespeare introduced a common error of his day – that snakes were slimy: but a cave seems to be a more likely habitat for a cobra than for a desert viper.

Virgil wrote of 'slippery serpents', Thomas Chatterton asserted that 'the slimy serpent swelters in its course', while Byron went to the extreme of stating the colour of the mythical slime: 'If like a snake she steals within your walls, Till the black slime betray her as she crawls'.

In olden times it was believed that the delicate forked tongue of a snake was its 'sting'. In the book of Job is written 'The viper's tongue shall slay him', but most Biblical writers speak correctly of the bites of serpents. Chaucer, however, attributed the death of Cleopatra to the 'sting' of a serpent, while Shakespeare in many of his plays referred to the 'sting' and 'double tongue' of the viper. Dryden went so far as to claim that 'A serpent shoots its sting'. Numerous writers, including Byron, describe the alleged faculty of snakes to fascinated their prey and John Wesley wrote that if a swallow pursuing its prey in the air should cast its eye upon a snake on the ground below, it would flutter over it 'in the utmost consternation, till sinking gradually lower and lower he at last drops into his mouth'.

## 12.5 Declining Amphibians

No myth, indeed a recurring theme in herpetological literature today, is the fact that amphibians, particularly frogs and toads, are declining rapidly in numbers throughout the world. This has been documented unequivocably by Stebbins and Cohen (1995), Beebee (1996) and others. One of the most disturbing aspects of recent reports of declines in the populations of amphibians around the world is that several have come from national parks, nature reserves and other protected and supposedly pristine environments. Many, but not all, of these reports have been anecdotal. A recent study near Yosemite National Park, California, however, provides clear evidence of a decline in a protected habitat. A transect in the Sierra Nevada paralleling one that was undertaken in the early 1900s showed that of the seven frog and toad species listed in the earlier survey, at least five have suffered serious decline. The red-legged frog (*Rana aurora*) has disappeared altogether, while another species, the yellow-legged frog (*R. muscosa*), now only survives in a few remnant populations. Previously it was the most abundant amphibian in the area. Only one species, the Pacific tree frog (*Hyla regilla*), remains nearly as widespread as it used to be 90 years ago, even though it has declined at elevations above 1220 m (4000 ft).

This situation is mirrored in several sites in the west coast states of the United States, where marked declines in toads (*Bufo* spp.) and frogs (*Rana* spp.) have been noted, while only *Hyla regilla* remains abundant. Possible local causes for these declines could include the introduction of predatory fishers, pollution and long-term drought. Bacterial and viral diseases may be catastrophic locally, but could only cause global decline under stress-related circumstances. The worldwide decline of frogs and toads is therefore still unexplained. It might possibly be due to the thinning of the ozone layer. Many amphibian eggs are vulnerable to ultra-violet radiation streaming through the upper atmosphere: UV radiation damages DNA, causing cell mutations and death. In many animals an enzyme can repair such damage, but some species of amphibians have very low levels of this enzyme and it is those species that are vanishing. Frogs are especially vulnerable on account of their specialized habitats and thin skins. Doubtless, combinations of factors are involved.

Global pollution could possibly be another cause of the catastrophe. Amphibians are by no means the only animals to be affected by man-made chemicals. Although reptiles are less sensitive, up to 80% of male alligators in one Florida lake are said to be suffering from abnormalities of the penis, while 75% of the alligator eggs studied there were found to be dead. It is believed that many insecticides may act like weak oestrogens and imitate female and other hormones. Lake Apopka in Florida was contaminated over a decade ago by a pesticide spill containing DDT from a chemical factory on its shores. This may have been responsible for the sexual abnormalities found among its alligators. Undoubtedly many European and Asian ponds and lakes are polluted to varying degrees as are the Great Lakes of North America. As we know, amphibians and reptiles play an important role in the nutrient flow of many food webs. Their decline might well herald as yet unknown ecological repercussions.

Both amphibians and, to a lesser extent, reptiles are indicators of environmental change and there is a correlation between species diversity and the diversity of resources (Pianka 1977). Habitat selection in snakes has been reviewed by Reinert (1993). Many human activities are destroying the environmental resources of both amphibians and reptiles. To protect them, their habitats must be protected. Many other things can be done too – for instance, wherever amphibians regularly cross roads to their breeding ponds tunnels should be constructed with drift fences leading to their entrances (Langton 1989). Conservation is discussed by Russell et al. (1992), Beebee (1996), and in several other books listed in the Bibliography.

## 12.6 The Role of the Naturalist

There are many aspects of herpetology that have not been discussed or are barely mentioned in this book. Among them are sensory physiology, histology, genetics and various aspects of morphology. Indeed, I have selected topics that are likely to be of greatest interest to the amateur naturalists for whom the book is primarily intended.

The human brain is the finest computer in existence and for many purposes a pencil and paper the most appropriate and convenient method of recording observations. There are infinitely more problems to be solved than have yet been thought about, let alone investigated. The creative pleasures of natural history frequently include making new observations, and it is amazing how much can be learned just by watching how animals behave when left to themselves. Furthermore, if a suitable research problem does not immediately suggest itself, the investigator can always fall back on measuring or weighing a number of specimens with a spring balance. A geometric plot of the results may reveal the number of age groups in the population, size differences between the sexes, and so on. With luck one can even observe predation or defence actually taking place. Most of our knowledge of these topics is based on inference, but raw observations may be the bases of interesting hypotheses.

One often reads of experiments that have been 'designed to prove' some hypothesis. This is quite misleading. It is important to remember that the scientist should not be trying to prove anything. A proper experiment is designed to test whether a particular hypothesis is correct or not. One has to keep an open mind, otherwise invalid conclusions may be reached.

In this book I have attempted to show how fascinating and diverse as objects of study the Amphibia and Reptilia prove themselves to be. In some respects the two classes are astonishingly similar; in others they are surprisingly different. The same is true of the various taxa within them. My object has been to present the view of an elderly professional zoologist with leanings towards natural history in a palatable form so that the intelligent amateur naturalist and student of herpetology may enjoy his or her recreation to an even greater extent than he or she already does.

# Bibliography

The selection of references to include in this bibliography has involved some difficulties because the literature pertinent to the subject is so vast. No attempt, therefore, has been made to list all the publications upon which the text has been based. To produce anything like a complete list would have rendered the book quite unwieldy. In general, therefore, reviews and advanced text have usually been cited in preference to the original research publications. Especially significant as sources of reference are the following:
Bellairs (1969), Bellairs and Cox (1976), Bradshaw (1986, 1997), Carpenter et al. (1994), Cloudsley-Thompson (1971), Cott (1961), Duellman and Trueb (1986), Dunson (1975a), Feder and Burggren (1992), Goin and Goin (1971), Gray (1968), Greene (1997), Heatwole and Taylor (1987), Hotton et al. (1986), Huey and Slatkin (1976), King (1996), Langton (1989), Lofts (1974a, 1976), McGowan (1991), Mayhew (1968), Moore (1964), Neill (1971), Noble (1931), Pianka (1986), Porter (1972), Romer (1966), Seigel and Collins (1993), Spellerberg (1973), Stebbins and Cohen (1995), Taylor and Guttman (1997), Templeton (1970), Thorpe et al. (1997), Vial (1973), Vitt and Pianka (1994), Whittow (1970), Zug (1993) and the volumes edited by Heatwole et al. in the Series: Amphibian biology and by Gans et al. in the Series: Biology of the Reptilia.

In addition, some recent publications not yet assumed into the general literature have been cited, as well as a number of research papers that are especially relevant to the topics of this book and which are less well known than the average.


Adler K (ed) (1992) Herpetology. Current research on the biology of amphibians and reptiles (Proc First World Congr Herpetol). Soc Study Amphib Rept, Oxford, Ohio, 245 pp

Alderton D (1988) Turtles and tortoises of the world. Blandford Press, London, 191 pp

Alexander R Mc N (1989) Dynamics of dinosaurs and other extinct giants. Columbia University Press, New York, 224 pp

Alvarez W (1997) *T. rex* and the crater of doom. Princeton University Press, Princeton, New Jersey, 185 pp

Alvarez L W, Alvarez F A, Michel H V (1980) Extra-terrestrial cause for the Cretaceous-Tertiary extinction. Science 208:1095–1108

Alvarez W, Kauffman E G, Surlyk F, Alvarez L W, Asaro F, Michel H V (1984) Impact theory of mass extinctions and the invertebrate fossil record. Science 223:1135–1141

Arak A (1983) Male-male competition and mate choice in anuran amphibians. In: Bateson PPG (ed) Mate choice. Cambridge University Press, Cambridge, pp 181–210

Arnold E N (1984) Evolutionary aspects of tail shedding in lizards and their relatives. J Nat Hist 18:127–169

Arnold E N (1988) Caudal autotomy as a defense. In: Gans C, Huey R B (eds) Biology of the Reptilia, vol 16. Ecology B. Defense and life history. Alan R Liss, New York, pp 235–273

Arnold S J (1977) The evolution of courtship behavior in New World salamanders with some comments on Old World salamandrids. In: Taylor D H, Gutterman S I (eds) The reproductive biology of amphibians. Plenum Press, New York, pp 141–183

Arnold S J (1993) Foraging theory and prey-size – predator-size relations in snakes. In: Seigel R A, Collins J T (eds) Snakes. Ecology and behavior. McGraw-Hill, New York, pp 87–115

Avery R A (1982) Field studies of body temperatures and thermoregulation. In: Gans C, Pough F H (eds) Biology of the Reptilia, vol 16. Ecology B Defense and life history. Alan R Liss, New York, pp 93–166

Avery R A (1996) Ecology of small reptile-grade Sauropsids. Symp Zool Soc Lond 69:225–237

Bagnara J T (1976) Colour change. In: Lofts B (ed) The physiology of Amphibia, vol III. Academic Press, London, pp 1–52

Bakker R (1987) The dinosaur heresies. A revolutionary view of dinosaurs. Longman Scientific & Technical, Harlow, Essex, 482 pp

Bakker R T (1971) Dinosaur phylogeny and the origin of mammals. Evolution 25:636–658

Bakker R T (1972) Anatomical and ecological evidence of endothermy in dinosaurs. Nature 238:81–85

Bartholomew G A (1982) Physiological control of body temperature. In: Gans C, Pough F H (eds) Biology of the Reptilia, vol 8. Physiology B. Academic Press, London, pp 167–211

Bates H W (1862) Contributions to an insect fauna of the Amazon valley. Lepidoptera: Heliconidae. Trans Linn Soc Lond 23:495–566

Bauer A M, Russell A P (1991) Pedal specialisations in dune-dwelling geckos. J Arid Environ 20:43–62

Beebee T J C (1995) Tadpole growth: is there an interference effect in nature? Herpetol J 5: 204–205

Beebee T J C (1996) Ecology and conservation of amphibians (Conservation Biology Series 7). Chapman & Hall, London, viii + 214 pp

Behrensmeyer A K, Damuth J P, Di Michele W A, Potts R, Sues H.-D, Wing S L (eds) (1992) Terrestrial ecosystems through time. Evolutionary paleoecology of terrestrial plants and animals. University of Chicago Press, Chicago, xx + 568 pp

Bellairs A (1969) The life of reptiles. Weidenfeld and Nicolson, London, vol I: xii + 282 pp, vol II: pp 283–590

Bellairs A d'A, Attridge J (1975) Reptiles (4th edn revised). Hutchinson University Library, London, 240 pp

Bellairs A d'A, Bryant S V (1985) Autotomy and regeneration in reptiles. In: Gans C, Billett F (eds) Biology of the Reptilia, vol I. Morphology A. Academic Press, New York, pp 301–410

Bellairs A, Carrington R (1966) The world of reptiles. Chatto & Windus, London, 153 pp

Bellairs A d'A, Cox C B (eds) (1976) Morphology and biology of reptiles. (Linnean Society Symposium Series No 3). Academic Press, London, xi + 290 pp

Bennett A F, Dawson W R (1976) Metabolism. In: Gans C, Dawson W R (eds) Biology of the Reptilia, vol 5. Physiology A. Academic Press, London, pp 127–223

Bennett A F, Ruben J A (1986) The metabolic and thermoregulatory status of therapsids. In: Hotton N, Maclean P D, Roth J J, Roth E C (eds) The ecology and biology of mammal-like reptiles. Smithsonian Institute Press, Washington, pp 207–218

Bentley P J (1976) Osmoregulation. In: Gans C, Dawson W R (eds) Biology of the Reptilia, vol 5. Physiology A. Academic Press, London, pp 365–412

Bentley P J, Schmidt-Nielsen K (1965) Permeability to water and sodium of the crocodilian *Caiman sclerops*. J Cell Comp Physiol 66:303–309

Bernard C (1878) Leçons sur les phénomènes de la vie commune aux animaux et aux végétaux. JB Baillière, Paris, xxxii + 404 pp

Berry J F, Shine R (1980) Sexual size dimorphism and sexual selection in turtles (order Testudines). Oecologia (Berl) 44:185–191

Birchard G F, Marcellini D (1996) Incubation time in reptilian eggs. J Zool (Lond) 240:621–635

Bogert C M (1960) The influence of sound on the behavior of amphibians and reptiles. In: Lanyon W F, Tavolga W N (eds) Animal sounds and communication. American Institute of Biological Sciences, Washington, DC, Publ No 7, pp 137–320

Borradaile L A (1923) The animal and its environment. Henry Frowde and Hodder & Stoughton, London, vii + 399 pp

Boutilier R G, Stiffler D F, Toews D P (1992) Exchange of respiratory gases, ions, and water in amphibious and aquatic amphibians. In: Feder M E, Burggren W W (eds) Environmental physiology of the amphibians. University of Chicago Press, Chicago, pp 81–124

Boyer D (1965) Ecology of the basking habit in turtles. Ecology 46:99–118

Bradshaw S D (1986) Ecophysiology of desert reptiles. Academic Press, London, xxv + 324 pp

Bradshaw S D (1997) Homeostasis in desert reptiles (Adaptations of desert organisms). Springer, Berlin Heidelberg New York, xi + 213 pp

Bramwell C D, Fellgett P B (1973) Thermal regulation in sail-lizards. Nature 242:203–205

Brattstrom B H (1970) Amphibia. In: Whittow G C (ed) Comparative physiology of thermoregulation, vol I. Invertebrates and nonmammalian vertebrates. Academic Press, New York, pp 135–166

Brattstrom B H (1974) The evolution of reptilian social behaviour. Am Zool 14:35–49

Brodie E D III, Brodie E D Jr (1990) Tetrodotoxin resistance in garter snakes: an evolutionary response of predators to dangerous prey. Evolution 44:651–659

Brown G W Jr (1964) The metabolism of Amphibia. In: Moore J A (ed) Physiology of the Amphibia, vol I. Academic Press, London, pp 1–98

Bustard H R (1967) Activity cycle and thermoregulation in the Australian gecko *Gehyra variegata*. Copeia 1967:753–758

Buxton P A (1923) Animal life in deserts. A study of the fauna in relation to the environment. Edward Arnold, London, xv + 176 pp

Caldwell J P (1996) The evolution of myrmecophagy and its correlates in poison frogs (family Dendrobatidae). J Zool (Lond) 240:75–101

Capranica R R (1976) The auditory system. In: Lofts B (ed) The physiology of Amphibia, vol III. Academic Press, London, pp 443–466

Carpenter K, Hirsh K F, Horner J R (eds) (1994) Dinosaur eggs and babies. Cambridge University Press, Cambridge, xiv + 372 pp

Carroll R L (1969a) Problems of the origin of reptiles. Biol Rev 44:393–432

Carroll R L (1969b) Origin of reptiles. In: Gans C, Bellairs A d'A, Parsons T S (eds) Biology of the Reptilia, vol I. Morphology A. Academic Press, London, pp 1–44

Carroll R L (1988) Vertebrate paleontology and evolution. WH Freeman, New York, xiv + 698 pp

Charig A (1976) Dinosaur monophyly and a new class of vertebrates. In: Bellairs A d'A, Cox C B (eds) Morphology and biology of reptiles (Linnean Society Symposium Series No 3). Academic Press, London, pp 65–104

Charig A (1979) A new look at the dinosaurs. Heinemann, London, 160 pp

Charig A J (1989) The Cretaceous-Tertiary boundary and the last of the dinosaurs. Philos Trans R Soc Lond B 325:387–400

Chiappe L M (1995) The first 85 million years of avian evolution. Nature 375:349–355

Christian K A, Tracy C R, Porter W P (1983) Seasonal shifts in body temperature and in the use of microhabitats by the Galapagos land iguana. Ecology 64:463–468

Clarke B T (1996) Small size in amphibians – its ecological and evolutionary implications. Symp Zool Soc Lond 69:201–224

Cloudsley-Thompson J L (1964) Diurnal rhythm of activity in the Nile crocodile. Anim Behav 12:98–100

Cloudsley-Thompson J L (1965) Rhythmic activity, temperature-tolerance, water-relations and mechanism of heat death in a tropical skink and gecko. J Zool (Lond) 146:55–69

Cloudsley-Thompson J L (1967) Diurnal rhythm, temperature and water relations of the African toad *Bufo regularis*. J Zool (Lond) 152:43–54

Cloudsley-Thompson J L (1969) Water-relations of the young Nile crocodile. Br J Herpetol 4:107–112

Cloudsley-Thompson J L (1970a) On the biology of the desert tortoise *Testuda sulcata* in Sudan. J Zool (Lond) 160:17–33

Cloudsley-Thompson J L (1970b) The significance of cutaneous respiration in *Bufo regularis* Reuss. Int J Biometeorol 14:361–364

Cloudsley-Thompson J L (1971) The temperature and water relations of reptiles. Merrow, Watford, Herts, 159 pp

Cloudsley-Thompson J L (1972) Temperature regulation in desert reptiles. Symp Zool Soc Lond 31:39–59

Cloudsley-Thompson J L (1974) Physiological thermoregulation in the spurred tortoise (*Testudo graeca* L.). J Nat Hist 8:577–587

Cloudsley-Thompson J L (1978a) Form and function in animals (Patterns of Progress 10). Meadowfield Press, Shildon Co Durham, 81 pp

Cloudsley-Thompson J L (1978b) Why the dinosaurs became extinct (Patterns of Progress 15). Meadowfield Press, Shildon Co Durham, 85 pp

Cloudsley-Thompson J L (1980a) Biological clocks. Their functions in nature. Weidenfeld and Nicolson, London, ix + 138 pp

Cloudsley-Thompson J L (1980b) Tooth and claw. Defensive strategies in the animal world. JM Dent, London, 252 pp

Cloudsley-Thompson J L (1981) Bionomics of the rainbow lizard *Agama agama* (L.) in eastern Nigeria during the dry season. J Arid Environ 4:235–245

Cloudsley-Thompson J L (1991) Ecophysiology of desert arthropods and reptiles (Adaptations of desert organisms). Springer, Berlin Heidelberg New York, x + 203 pp

Cloudsley-Thompson J L (1993) When the going gets hot, the tortoise gets frothy. Nat Hist 102(8):33–35

Cloudsley-Thompson J L (1994) Predation and defence amongst reptiles. R & A, Taunton, Somerset, viii + 138 pp

Cloudsley-Thompson J L (1996) Biotic interactions in arid lands (Adaptations of desert organisms). Springer, Berlin Heidelberg New York, xi + 208 pp

Cogger H G (1975) Sea snakes of Australia and New Guinea. In: Dunson W A (ed) The biology of sea snakes. University Park Press, Baltimore, pp 59–139

Colbert E H (1962) Dinosaurs. Their discovery and their world. Hutchinson, London, 288 pp

Colbert E H (1965) The age of reptiles. Weidenfeld and Nicolson, London, xiv + 228 pp
Colbert E H (1980) Evolution of the vertebrates. A history of the backboned animals through time, 3rd edn. John Wiley, New York, xvi + 510 pp
Colbert E H (1986) Therapsids in Pangaea and their contemporaries and competitors. In: Hotton N III, Maclean P D, Roth J J, Roth E C (eds) The ecology and biology of mammal-like reptiles. Smithsonian Institute Press, Washington, pp 133–145
Collette B B (1961) Correlations between ecology and morphology in anoline lizards from Havana, Cuba and southern Florida. Bull Mus Comp Zool Harv College 125:137–162
Constantinou C, Cloudsley-Thompson J L (1985) The circadian rhythm of locomotory activity in the desert lizard *Acanthodactylus schmidti*. J Interdiscip Cycle Res 16:107–111
Cooper W E Jr, Greenberg N (1992) Reptilian coloration and behavior. In: Gans C, Crews D (eds) Biology of the Reptilia, vol 18. Hormones, brain and behavior. University of Chicago Press, Chicago, pp 298–422
Cott H B (1926) Observations on the life-habits of some batrachians and reptiles from the lower Amazon, and a note on some mammals from Marajo Island. Proc Zool Soc Lond 1926:1159–1178
Cott H B (1940) Adaptive coloration in animals. Methuen, London, xxxii + 508 pp
Cott H B (1961) Scientific results of an inquiry into the ecology and economic status of the Nile crocodile (*Crocodilus niloticus*) in Uganda and Northern Rhodesia. Trans Zool Soc Lond 29(4):211–337
Croft L R (1982) The last dinosaurs. A new look at the extinction of the dinosaurs. Elmwood, Chorley, Lancs 80 pp
Crump M L (1992) Cannibalism in amphibians. In: Elgar M A, Crespi B J (eds) Cannibalism, ecology and evolution among diverse taxa. Oxford University Press, Oxford, pp 256–276
Daltry J C, Wüster W, Thorpe R S (1996) Diet and snake venom evolution. Nature 379: 537–540
Dantzler W H (1976) Renal function (with special emphasis on nitrogen excretion). In: Gans C, Dawson W R (eds) Biology of the Reptilia, vol 5. Physiology A. Academic Press, London, pp 447–503
Darevsky I S, Kupriyanova L A, Uzzell T (1985) Parthenogenesis in reptiles. In: Gans C, Billett F (eds) Biology of the Reptilia, vol 1. Morphology A. Academic Press, London, pp 411–526
Desmond A J (1975) The hot-blooded dinosaurs. A revolution in palaeontology. Blond & Briggs, London, 238 pp
Deyrup I J (1964) Water balance and kidney. In: Moore J A (ed) Physiology of the Amphibia, vol 1. Academic Press, New York, pp 251–328
Diefenbach C O da C (1973) Integumentary permeability to water in *Caiman crocodilus* and *Crocodylus niloticus* (Crocodilia: Reptilia). Physiol Zool 46:72–78
Diesel R, Bäurle G, Vogel P (1995) Cave breeding and froglet transport: a novel pattern of anuran brood care in the Jamaican frog, *Eleutherodactylus cundalli*. Copeia 1995:354–360
Dobkin D S, Gettinger R D (1985) Thermal aspects of anuran foam nests. J Herpetol 19:271–275
Done B S, Heatwole H (1977) Social behavior of some Australian skinks. Copeia 1977:419–430
Downie J R (1988) Functions of the foam in the foam-nesting leptodactylid *Physalaemus pustulosus*. Herpetol J 1:302–307
Drewes R C, Altig R (1996) Anuran egg predation and hetero-cannibalism in a breeding community of East African frogs. Trop Zool 9:333–347
Duellman W E, Trueb L (1986) Biology of amphibians. McGraw-Hill, New York, xvii + 670 pp

Dunham A E, Morin P J, Wilbur H M (1988) Methods for the study of reptile populations. In: Gans C, Huey R B (eds) Biology of the Reptilia, vol 16. Ecology B. Defence and life history. Alan R Liss, New York, pp 331–386

Dunson W A (ed) (1975a) The biology of sea snakes. University Park Press, Baltimore, x + 530 pp

Dunson W A (1975b) Adaptations of sea sakes. In: Dunson W A (ed) The biology of sea snakes. University Park Press, Baltimore, pp 13–19

Dunson W A (1975c) Salt and water balance of sea snakes. In: Dunson W A (ed) The biology of sea snakes. University Park Press, Baltimore, pp 329–353

Dunson W A (1976) Salt glands in reptiles. In: Gans C, Dawson W R (eds) Biology of the Reptilia, vol 5. Physiology A. Academic Press, London, pp 413–445

Duval D, Guillette L J Jr, Jones R E (1982) Environmental control of reptilian reproductive cycles. In: Gans C, Pough F H (eds) Biology of the Reptlia, vol 16. Ecology B Defence and life history. Alan R Liss, New York, pp 201–231

Edmunds M (1974) Defence in animals. A survey of anti-predator defences. Longman, Harlow, Essex, xvii + 357 pp

Elkan E (1968) Mucopolysaccharides in the anuran defence against desiccation. J Zool (Lond) 155:19–53

Elkan E (1976) Ground substance: an anuran defense against desiccation. In: Lofts G (ed) The physiology of Amphibia, vol III. Academic Press, London, pp 101–110

Elliott W B (1978) Chemistry and immunology of reptile venoms. In: Gans C, Gans K A (eds) Biology of the Reptilia, vol 8. Physiology B. Academic Press, London, pp 163–426

Engelmann W-E, Obst F J (1984) Snakes. Biology, behavior and relationship to man. Croom Helm, London, 222 pp

Feder M E, Burggren W W (eds) (1992) Environmental physiology of the amphibians. University of Chicago Press, Chicago, viii + 646 pp

Feduccia A (1980) The age of birds. Harvard University Press, Cambridge Massachusetts, ix + 196 pp

Feduccia A (1996) The origin and evolution of birds. Yale University Press, London, x + 420 pp

Fenton M B, Licht L E (1990) Why rattle snake? J Herpetol 24:274–279

Fitch H S, Greene H W (1965) Breeding cycle in the ground skink *Lygosoma laterale*. Univ Kans Pub Mus Nat Hist 15:565–575

Fogden M, Fogden P (1988) Tropical frogs and common toads. On water and on land. World Mag 13:1–8

Frazer D (1983) Reptiles and amphibians in Britain (The New Naturalist). Collins, London, 254 pp

Frazer JFD (1973) Amphibians (The Wykeham Science Series). Wykeham Publications, London, vi + 122 pp

Gadow H (1901) Amphibia and reptiles (The Cambridge Natural History, vol VIII). Macmillan, London, xiii + 669 pp

Gans C (1961) The feeding mechanism of snakes and its possible evolution. Am Zool 1:217–227

Gans C (1962) Terrestrial locomotion without limbs. Am Zool 2:167–182

Gans C (1970) How snakes move. Sci Am 222(6):82–96

Gans C (1978) Reptilian venoms: some evoltionary considerations. In: Gans C, Gans K A (eds) Biology of the Reptilia, vol 5. Physiology A. Academic Press, London, pp 1–42

Gans C (1983) Snake feeding strategies and adptations-conclusion and prognoses. Am Zool 23:455–460

Gans C, Billett F (eds) (1985) Biology of the Reptilia, vol 15. Development B. x + 731 pp

Gans C, Crews D (eds) (1992) Biology of the Reptilia, vol 18. Physiology E. Hormones, brain and behavior. University of Chicago Press, Chicago, xi + 564 pp

Gans C, Dawson W R (eds) (1976) Biology of the Reptilia vol 5. Physiology A. Academic Press, London, xv + 556 pp

Gans C, Gans K A (eds) (1978) Biology of the Reptilia, vol 8. Physiology B. Academic Press, London, xiii + 782 pp

Gans C, Gorniak G C (1982) How does the toad flip its tongue? Test of two hypotheses. Science 216:1335–1337

Gans C, Huey R B (eds) (1988) Biology of the Reptilia, vol 16. Ecology B. Defense and life history. Alan R Liss, New York, xi + 659 pp

Gans C, Pough F H (eds) (1982) Biology of the Reptilia, vol 13. Physiology D. Physiological ecology. Academic Press, London, xiii + 345 pp

Gans C, Tinkle D W (eds) (1977) Biology of the Reptilia, vol 7. Ecology and behaviour A. Academic Press, London, xvi + 720 pp

Gans C, Bellairs A d'A, Parsons T S (eds) (1969) Biology of the Reptilia, vol 1. Morphology A. Academic Press, London, xv + 373 pp

Gans C, Billett F, Maderson P F A (eds) (1985) Biology of the Reptilia, vol 14. Development A. John Wiley, New York, xii + 763 pp

Garland T Jr. (1994) Phylogenetic analyses of lizard endurance capacity in relation to body size and body temperature. In: Vitt L C, Pianka E R (eds) Lizard ecology. Historical and experimental perspectives. Princeton University Press, Princeton, New Jersey, pp 237–259

Girgis S (1961) Aquatic respiration in the common Nile turtle *Trionyx triunguis* (Forskål). Comp Biochem Physiol 3:206–217

Goin C J, Goin O B, Zug G R (1978) Introduction to herpetology, 3rd edn. WH Freeman, San Francisco, 353 pp

Gorzula S (1977) Foam nesting in leptodactylids: a possible function. Br J Herpetol 5: 657–659

Gourley E V (1972) Circadian activity rhythm of the gopher tortoise (*Gopherus polyphemus*). Anim Behav 20:13–20

Graham T E, Hutchison V H (1979) Turtle diel activity. Response to different regimes of temperature and photoperiod. Comp Biochem Physiol 63A:299–305

Gray J (1968) Animal locomotion. Weidenfeld and Nicolson, London, xi + 479 pp

Greene H W (1988) Antipredator mechanisms in reptiles. In: Gans, Huey (eds) Biology of the Reptilia, vol 16. Ecology B. Defense and life history. Alan R Liss, New York, pp 1–152

Greene H W (1997) Snakes. The evolution of mystery in nature. University of California Press, Berkeley, xiii + 351 pp

Greene H W, McDiarmid R W (1981) Coral snake mimicry: does it occur? Science 213: 1207–1212

Gregory P T (1982) Reptilian hibernation. In: Gans C, Pough F H (eds) Biology of the Reptilia, vol 13. Physiology D. Physiological ecology. Academic Press, London, pp 53–154

Griffith H (1994) Body elongation and decreased reproductive output within a restricted clade of lizards (Reptilia: Scinidae). J Zool (Lond) 233:541–550

Griffiths R (1996) Newts and salamanders of Europe. T & AD Poyser, London, x + 188 pp

Griffiths R A, Denton J, Wong AL-C (1993) The effect of food level on competition in tadpoles: interference mediated by protothecan algae. J Anim Ecol 62:274–279

Griffiths R A, Edgar P W, Wong AL-C (1991) Interspecific competition in tadpoles: growth inhibition and growth retrieval in natterjack toads *Bufo calamita*. J Anim Ecol 60:1065–1076

Gross M R, Shine R (1981) Parental care and mode of fertilisation in ectothermic vertebrates. Evolution 35:775–793

Guggisberg C A W (1972) Crocodiles. Their natural history, folklore and conservation. David & Charles Publishers, Newton Abbot, x + 204 pp

Guillette L J Jr, Hotton N III (1986) The evolution of mammalian reproductive characteristics in therapsid reptiles. In: Hotton N, Maclean P D, Roth J J, Roth E C (eds) The ecology and biology of mammal-like reptiles. Smithsonian Institute Press, Washington, pp 239–250

Hailey A, Coulson I M (1996a) Temperature and the tropical tortoise *Kinixys spekii*: constraints on activity level and body temperature. J Zool (Lond) 240:523–536

Hailey A, Coulson I M (1996b) Temperature and the tropical tortoise *Kinixys spekii*: tests of thermoregulation. J Zool (Lond) 240:537–549

Hallam A, Wignall P B (1997) Mass extinctions and their aftermath. Oxford University Press, Oxford, 320 pp

Halliday T R (1975) An observational and experimental study of sexual behaviour in the smooth newt *Triturus vulgaris* (Amphibia: Salamandridae). Anim Behav 23:291–322

Halliday T R (1992) Sexual selection in amphibians and reptiles: theoretical issues and new directions. In: Adler K (ed) Herpetology. Current research on the biology of amphibians and reptiles. Soc Study Amphib Rept, Oxford, Ohio, pp 81–95

Halstead L B, Halstead J (1981) Dinosaurs. Blandford Press, Poole, 170 pp

Harris V A (1964) The life of the rainbow lizard (Hutchinson Tropical Monographs). Hutchinson, London, 174 pp

Heath J E (1965) Temperature regulation and diurnal activity in horned lizards. Univ Calif Publ Zool 64:97–136

Heath J E (1966) Venous shunts in the cephalic sinuses of horned lizards. Physiol Zool 39: 30–35

Heatwole H (1970) Thermal ecology of the desert dragon *Amphibolurus inermis*. Ecol Monogr 40:425–457

Heatwole H (1975) Predation on sea snakes. In: Dunson W A (ed) The biology of sea snakes. University Park Press, Baltimore, pp 233–249

Heatwole H (1976) Reptile ecology. University of Queensland Press, St Lucia, Queensland, xviii + 178 pp

Heatwole H (1984) Adaptations of amphibians to aridity. In: Cogger H G, Cameron E E (eds) Arid Australia. Australian Museum, Sydney, pp 177–222

Heatwole H (1987) Sea snakes. University of New South Wales Press, Kensington NSW, viii + 85 pp

Heatwole H, Barthalmus G T (eds) (1994) Amphibian biology, vol 1. The integument. Surrey Beatty, Chipping Norton NSW, xii + 418 pp

Heatwole H, Seymour R (1975) Diving physiology. In: Dunson W A (ed) The biology of sea snakes. University Park Press, Baltimore, pp 289–327

Heatwole H, Sullivan B K (eds) (1995) Amphibian biology, vol 2. Social behaviour. Surrey Beatty, Chipping Norton NSW, xii + 290 pp

Heatwole H, Taylor J (1987) Ecology of reptiles. Surrey Beatty, Chipping Norton NSW, xvi + 325

Heinrich B (1977) Why have some animals evolved to regulate a high body temperature? Am Nat 111:623–640

Herman C A (1992) Endocrinology. In: Feder M E, Burggren W W (eds) Environmental physiology of the amphibians. University of Chicago Press, Chicago, pp 40–57

Hertz P E, Huey R B, Nevo E (1982) Fight versus flight: body temperature influences defensive responses of lizards. Anim Behav 30:676–679

Heuvelmans B (1959) On the track of unknown animals (Transl Garnett R). Rupert Hart-Davies, London, 558 pp

Heuvelmans B (1968) In the wake of the sea-serpents (Transel Garnett R). Rupert Hart-Davis, London, 645 pp

Higginbotham A C (1939) Studies in amphibian activity 1. Preliminary report on the rhythmic activity of *Bufo americanus* Holbrook and *Bufo fowleri* Hinckley. Ecology 20:58–70

Hock R J (1964) Terrestrial animals in cold: reptiles. In: Dill D B (ed) Handbook of physiology, Sect 4: Adaptation to the environment. American Physiological Society, Washington DC, pp 357–359

Hotton N III, Maclean P D, Roth J J, Roth E C (eds) (1986) The ecology and biology of mammal-like reptiles. Smithsonian Institute Press, Washington, x + 326 pp

Howard R R, Brodie E D Jr (1973) Experimental study of Batesian mimicry in the salamanders *Plethodon jordani* and *Desmognathus ochrophaeus*. Am Midl Nat 60:38–46

Huey, R B (1982) Temperature physiology and the ecology of reptiles. In: Gans C, Pough F H (eds) Biology of the Reptilia, vol 13. Physiology D Physiological ecology. Academic Press, London, pp 25–91

Huey R B, Pianka E R (1981) Ecological consequences of foraging mode. Ecology 62:991–999

Huey R B, Slatkin M (1976) Cost and benefits of lizard thermoregulation. Q Rev Biol 51:363–384

Huey R B, Pianka R, Schoener T W (eds) (1983) Lizard ecology. Studies of a model organism. Harvard University Press, Cambridge, Massachusetts, viii + 501 pp

Hutchison V H, Dupré R K (1992) Thermoregulation. In: Feder M E, Burggren W W (eds) Environmental physiology of amphibians. University of Chicago Press, Chicago, pp 206–249

Hutchison V H, Dowling H G, Vinegar A (1966) Thermoregulation in a brooding female Indian python *Python molurus bivittatus*. Science 151:694–696

Jackson K, Butler D G, Brooks D R (1996) Habitat and phylogeny influence salinity discrimination in crocodilians: implications for osmoregulatory physiology and historical biogeography. Biol J Linn Soc 58:371–383

Jaeger R G, Rubin A M (1982) Foraging tactics of a terrestrial salamander: judging prey profitability. J Anim Ecol 51:167–176

Jaegar R G, Schwarz J K (1991) Gradational threat postures by the red-backed salamander. J Herpertol 25:112–114

Jepsen G L (1964) Riddles of the terrible lizards. Am Sci 52:227–246

King G (1996) Reptiles and herbivory. Chapman & Hall, London, vii + 160 pp

Kobelt F, Linsenmair K E (1992) Adaptations of the reed frog *Hyperolius viridiflavus* (Amphibia: Anura: Hyperoliidae) to its arid environment. VI. The iridophores in the skin as radiation reflectors. J Comp Physiol (B) 162:314–326

Kochva E (1978) Oral glands of the Reptilia. In: Gans C, Gans C A (eds) Biology of the Reptilia, vol 8. Physiology B. Academic Press, London, pp 43–161

Kochva E, Nakar O, Ovadia M (1983) Venom toxins: plausible evolution from digestive enzymes. Am Zool 23:427–430
Kurtén B (1968) The age of the dinosaurs. Weidenfeld & Nicolson, London, 255 pp
Lack D (1939) The display of the blackcock. Brit Birds 32:290–303
Lambert D (1992) Dinosaur data book. Facts on File – British Museum (Natural History), New York, 320 pp
Lang J W (1989) Social behavior. In: Ross C A, Garnett S (eds) Crocodiles and alligators. Merehurst Press, London, pp 102–117
Langton T E S (ed) (1989) Amphibians and roads. Proc Toad Tunnel Conf, Rensburg, 7–8 January 1989. ACO Polymer Products, Shefford, Bedfordshire, 202 pp
Larsen L O (1976) Physiology of molting. In: Lofts B (ed) The physiology of Amphibia, vol III. Academic Press, London, pp 53–100
Larsen L O (1992) Feeding and digestion. In: Feder M E, Burggren W W (eds) Environmental physiology of the amphibians. University of Chicago Press, Chicago, pp 378–394
Lee M S Y (1996) Correlated progression and the origin of turtles. Nature 379:812–815
Lee A K, Mercer E H (1967) Cocoon surrounding desert-dwelling frogs. Science 157:87–88
Legler J M (1960) Natural history of the ornate box turtle *Terrapene ornata ornata* Agassiz. Univ Kans Publ Mus Nat Hist 11:527–669
Lehn W H (1979) Atmospheric refraction and lake monsters. Science 205:183–185
Licht P (1972) Problems in experimentation on timing mechanisms for annual physiological cycles in reptiles. In: South F E, Hannon J P, Willis J R, Pengelley E T, Alpert N R (eds) Hibernation and hypothermia, perspectives and challenges. Elsevier, Amsterdam, pp 681–711
Lillywhite H B, Henderson R W (1993) Behavioral and functional ecology of arboreal snakes. In: Seigel R A, Collins J T (eds) Snakes. Ecology and behavior. McGraw Hill, New York, pp 1–48
Lillywhite H B, Maderson P F A (1982) Skin structure and permeability. In: Gans C, Pough F H (eds) Biology of the Reptilia, vol 13. Physiology D. Physiological ecology. Academic Press, London, pp 397–442
Little C (1983) The colonisation of land. Origins and adaptations of terrestrial animals. Cambridge University Press, Cambridge, 290 pp
Lofts B (ed) (1974a) The physiology of Amphibia, vol II. Academic Press, London, xi + 592 pp
Lofts B (1974b) Reproduction. In: Lofts B (ed) The physiology of Amphibia, vol II. Academic Press, London, pp 107–218
Lofts B (ed) (1976) The physiology of Amphibia, vol III. Academic Press, London, xiv + 644 pp
Lohmann K J, Lohmann C M F (1996) Detection of magnetic field intensity by sea turtles. Nature 380:59–61
Louw G N (1972) The role of advective fog in the water economy of certain Namib desert animals. Symp Zool Soc Lond 31:297–314
Loveridge, J P (1984) Thomoregulation in the Nile crocodile, *Crocodylus niloticus*. Symp Zool Soc Lond 52:443–467
Loveridge J P, Withers P C (1981) Metabolism and water balance of active and cocooned African bullfrogs *Phxicaphalus adsperus*. Physiol Zool 54:203–214
Lucas J R, Howard R D, Palmer J G (1996) Callers and satellites: chorus behaviour in anurans as a stochastic dynamic game. Anim Behav 51:501–518
Luke C (1986) Convergent evolution of lizard toe fringes. Biol J Linn Soc 27:1–16

Macartney J M, Gregory P T, Larsen K W (1988) A tabular survey of data on movements and home ranges of snakes. J Herpetol 22:61–73
Mackal R P (1983) Searching for hidden animals. Cadogan Books, London, xxiv + 294
Mahoney J J, Hutchison V H (1960) Photoperiod acclimation and 24-hour variations in the critical thermal maxima of a tropical and temperate frog. Oecologia (Berl) 2:143–161
Martins E P (1994) Phylogenetic perspectives on the evolution of lizard territoriality. In: Vitt L J, Pianka E R (eds) Lizard ecology. Historial and experimental perspectives. Princeton University Press, Princeton, New Jersey, pp 117–144
Martins M (1989) Deimatic behavior in *Pleurodema brachyops*. J Herpetol 23:305–307
Mattison C (1987) Frogs and toads of the world. Blandford Press, Poole, Dorset, 191 pp
Mautz W J (1982) Patterns of evaporative water loss. In: Gans C, Pough F H (eds) Biology of the Reptilia, vol 13. Physiology D. Academic Press, London, pp 443–481
May R M (1991) How many species? Philos Trans R Soc Lond (B) 330:293–304
Mayhew W W (1965) Adaptations of the amphibian *Scaphiopus couchi*, to desert conditions. Am Midl Nat 74:95–109
Mayhew W W (1968) Biology of desert amphibians and reptiles. In: Brown G W Jr (ed) Desert biology. Special topics on the physical and biological aspects of arid regions, vol I. Academic Press, New York, pp 195–356
McGowan C (1991) Dinosaurs, Spitfires and sea dragons. Harvard University Press, Cambridge, Massachusetts, x + 365 pp
McNab B K (1978) The evolution of endothermy in the phylogeny of mammals. Am Nat 112: 1–21
Mertens R (1966) Das problem der mimicry bei Korallenschlangen. Zool Jahr Syst 84: 541–576
Miller C M (1944) Ecological relations and adaptations of the limbless lizards of the genus *Anniella*. Ecol Monogr 14:271–289
Milne A (1991) The fate of the dinosaurs. New perspectives in evolution and extinction. Prism Press, Bridport, Dorset, ix + 301 pp
Minnich J E (1982) The use of water. In: Gans C, Pough F H (eds) Biology of the Reptilia, vol 13. Physiology D. Physiological ecology. Academic Press, London, pp 325–395
Minton S A (1975) Geographic distribution of sea snakes. In: Dunson W A (ed) The biology of sea snakes. University Park Press, Baltimore, pp 21–31
Minton S A Jr, Minton M R (1971) Venomous reptiles. George Allen & Unwin, London, xii + 274 pp
Moore J A (ed) (1964) Physiology of the Amphibia, vol I. Academic Press, New York, xii + 654 pp
Mosauer W (1932) Adaptive convergence in the sand reptiles of the Sahara and of California. Copeia 1932:72–78
Müller F (1879) *Ituma* ad *Thyridia*; a remarkable case of mimicry in butterflies (transl Mendola R). Proc Entomol Soc Lond xx–xxix
Myers C W, Daly J W (1983) Dart-poison frogs. Sci Am 248(2):120–133
Nagy K A (1982) Field studies of water relations. In: Gans C, Pough F H (eds) Biology of the Reptilia, vol 13. Physiology D. Physiological ecology. Academic Press, London, pp 483–501
Neill W T (1971) The last of the ruling reptiles. Alligators, crocodiles and their kin. Columbia University Press, New York, xvii + 486 pp
Neill W T (1974) Reptiles and amphibians in the service of man. Pegasus, New York, ix + 248 pp

Nishikawa K C, Service P M (1988) A fluorescent marking technique for recognition of terrestrial salamanders. J Herpetol 22:351–353

Noble G K (1931) The biology of Amphibia. McGraw Hill, New York, 577 pp (Reprinted 1954. Dover Publications, New York)

Noble G K (1937) The sense organs involved in the courtship of *Storeria*, *Thamnophis* and other snakes. Bull Am Mus Nat Hist 73:673–725

Norman D (1985) The illustrated encyclopedia of dinosaurs. Salamander Books, London, 208 pp

Norman D (1991) Dinosaur! Boxtree, London, 288 pp

Norrell M A, Clark J M, Chiappe L M, Dashzveg D (1995) A nestling dinosaur. Nature 378: 774–776

Olson EC (1976) The exploitation of land by early tetrapods. In: Bellairs A d'A, Cox C B (eds) Morphology and biology of reptiles (Linnean Society Symposium Series No 3) Academic Press, London, pp 1–30

Olson E C (1986) Relationships and ecology of the early therapsids and their predecessors. In: Hotton N III, Maclean P D, Roth J J, Roth E C (eds) The ecology and biology of mammal-like reptiles. Smithsonian Institute Press, Washington, pp 47–60

Olsson M, Gullberg A, Tegelström H (1997) Determinants of breeding dispersal in the sand lizard *Lacerta agilis* (Reptilia, Squamata). Biol J Linn Soc 60:242–256

Ostrom J H (1976) *Archaeopteryx* and the origin of birds. Biol J Linn Soc 8:91–182

Packard G C, Packard M J (1988) The physiological ecology of reptilian eggs and embryos. In: Gans C, Huey RB (eds) Biology of the Reptilia, vol 16. Ecology B. Defense and life history. Alan R Liss, New York, pp 523–605

Parker H W (1977) Snakes: a natural history, 2nd edn. (Revised and enlarged by AGC Grandison). British Museum (Natural History) and Cornell University Press, London, Ithaca, 108 pp

Paul G S (1994a) Dinosaur reproduction in the fast lane: implications for size, success, and extinction. In: Carpenter K, Hirsch K F, Horner J R (eds) Dinosaur eggs and babies. Cambridge University Press, Cambridge, pp 244–255

Paul G S (1994b) Thermal environments of dinosaur nestlings: implications for endothermy and insulation. In: Carpenter K, Hirsch K, Horner J R (eds) Dinosaur eggs and babies. Cambridge University Press, Cambridge, pp 279–287

Peterson C R, Gibson A R, Dorcas M E (1993) Snake thermal ecology: the causes and consequences of body-temperature variation. In: Seigel R A, Collins J T (eds) Snakes. Ecology and behavior. McGraw-Hill, New York, pp 241–314

Phillips J B (1987) Laboratory studies of homing orientation in the eastern red-spotted newt *Notophthalmus viridescens*. J Exp Biol 131:215–229

Pianka E R (1977) Reptilian species diversity. In: Gans C, Tinkle D W (eds) Biology of the Reptilia, vol 7. Ecology and behavior A. Academic Press, London, pp 1–34

Pianka E R (1986) Ecology and natural history of desert lizards. Princeton University Press, Princeton, New Jersey, x + 201 pp

Pinder A W, Storey K B, Uitsch G R (1992) Estivation and hibernation. In: Feder M E, Burggren W W (eds) Environmental physiology of the amphibians. University of Chicago Press, Chicago, pp 250–274

Pooley A C, Hines T, Shield J (1989) Attacks on humans. In: Ross C A, Garnett S (eds) Crocodiles and alligators. Merehurst Press, London, pp 172–187

Pope C H (1955) The reptile world (Borzoi Books). Alfred A Knopf, New York, xxv + 325 + xiii pp

Porter K R (1972) Herpetology. W B Saunders, Philadelphia, xi + 524 pp
Pough F H (1969) The morphology of undersand respiration in reptiles. Herpetologia 25: 216–223
Pough F H (1973) Lizard energetics and diet. Ecology 54:837–844
Pough F H (1983) Feeding mechanisms, body size, and the ecology and evolution of snakes. Am Zool 23:339–342
Pough F H (1988) Mimicry and related phenomena. In: Gans C, Huey K B (eds) Biology of the Reptilia, vol 16. Ecology B. Defense and life history. Alan R Liss, New York, pp 153–234
Pough F H, Groves J D (1983) Specializations of the body form and food habits of snakes. Am Zool 23:443–454
Punzo F (1982) Tail autotomy and running speed in the lizards *Cophosaurus texana* and *Uma notata*. J Herpetol 16:329–331
Rahn H (1966) Gas transport from the external environment to the cell. In: de Reuck A V S, Porter R (eds) Development of the lung. CIBA Foundation Symposium. Churchill, London, pp 3–23
Reinert H K (1993) Habitat selection in snakes. In: Seigel R A, Collins J T (eds) Snakes. Ecology and behavior. McGraw-Hill, New York, pp 201–240
Richards C M (1962) The control of tadpole growth by alga-like cells. Physiol Zool 35:285–296
Ricqlès A de (1975) Les premiers vertébrés volants. Recherche 6:608–617
Ridley M (1978) Parental care. Anim Behav 26:904–932
Riedesel M L, Cloudsley-Thompson J A, Cloudsley-Thompson J L (1971) Evaporative thermoregulation in turtles. Physiol Zool 44:28–32
Rodda G H, Phillips J B (1992) Navigational systems develop along similar lines in amphibians, reptiles and birds. Ethol Ecol Evol 4:43–51
Röll B (1995) Epidermal fine structure of the toe tips of *Sphaerodactylus cinereus* (Reptilia, Gekkonidae). J Zool (Lond) 235:289–300
Romer A S (1948) Relative growth in pelycosaurian reptiles. Spec Publ Royal Society of South Africa. R Broom Commemoration Vol, pp 45–55
Romer A S (1966) Vertebrate paleontology (3rd edn). University of Chicago Press, Chicago, 468 pp
Romer A S (1968) The procession of life. Weidenfeld and Nicolson, London, viii + 323 pp
Romer A S, Parsons T S (1986) The vertebrate body (6th edn). Saunders College Publishing, Philadelphia, vii + 679 pp
Ross C A, Garnett S (eds) (1989) Crocodiles and alligators. Merehurst Press, London, 240 pp
Roze J A (1996) Coral snakes of the Americas. Biology, identification, and venoms. Krieger Publishing, Malabar, Florida, xii + 328 pp
Ruben J A (1986) Therapsids and their environment, a summary. In: Hotton N III, Maclean P D, Roth J J, Roth E C (eds) The ecology and biology of mammal-like reptiles. Smithsonian Institution Press, Washington, pp 307–312
Rubinoff I, Kropach C (1970) Differential reactions of Atlantic and Pacific predators to sea snakes. Nature 228:1288–1290
Russell A M, Carr J L, Swingland I R, Werner T B, Mast R B (1992) Conservation of amphibians and reptiles. In: Adler K (ed) Herpetology. Current research on the biology of amphibians and reptiles (Proc First Congr Herpetol). Soc Stud Amphib Rept, Oxford, Ohio, pp 59–80
Salthe S N, Mecham J S (1974) Reproductive and courtship patterns. In: Lofts B (ed) The physiology of Amphibia, vol II. Academic Press, London, pp 309–521

Savile D B O (1962) Gliding and flight in the vertebrates. Am Zool 2:161–166

Schiøtz A (1973) Evolution of anuran mating calls: ecological aspects. In: Vial J L (ed) Evolutionary biology of the anurans. University of Missouri Press, Columbia, Missouri, pp 311–319

Schmalhausen I I (1968) The origin of terrestrial vertebrates. (Transl Kelso L) Academic Press, New York, xxi + 314 pp

Schmid W D (1968) Natural variations in nitrogenous excretion of amphibians from different habitats. Ecology 49:180–185

Schmidt-Nielsen K, Dawson W R (1964) Terrestrial animals in dry heat: desert reptiles. In: Dill D B (ed) Handbook of physiology. Sect 4: Adaptation to the environment. American Physiological Society, Washington DC, pp 467–480

Secor S, Diamond J (1995) Adaptive responses to feeding in Burmese pythons: pay before pumping. J Exp Biol 198:1313–1325

Seigel R A, Collins J T (eds) (1993) Snakes. Ecology and behavior. McGraw-Hill, New York, xvi + 415 pp

Seymour R S (1982) Physiological adaptations to aquatic life. In: Gans C, Pough F H (eds) Biology of the Reptilia, vol 13. Physiology D. Physiological ecology. Academic Press, London, pp 1–51

Shine R (1985) The evolution of viviparity in reptiles: an ecological analysis. In Gans C, Billatt F, Maderson P F A (eds) Biology of the Reptilia, vol 14. Development A. John Wiley, New York, pp 605–694

Shine R (1988) Parental care in reptiles. In: Gans C, Huey R B (eds) Biology of the Reptilia, vol 16. Ecology B. Defence and life history. Alan R Liss, New York, pp 275–329

Shine R (1995) A new hypothesis for the evolution of viviparity in reptiles. Am Nat 145: 809–823

Shine R, Bull J J (1979) The evolution of live-bearing in lizards and snakes. Am Nat 113: 905–923

Shoemaker V H, Hillman S S, Hillyard S D, Jackson D C, McClanahan L L, Withers P C, Wygoda M L (1992) Exchange of water, ions, and respiratory gases in terrestrial amphibians. In: Feder M E, Burggren W W (eds) Environmental physiology of the amphibians. University of Chicago Press, Chicago, pp 125–150

Shuker K P N (1995) In search of prehistoric survivors. Do giant 'extinct' creatures still exist? Blandford, London, 192 pp

Sih A, Moore R D (1993) Delayed hatching of salamander eggs in response to enhanced larval predation risk. Am Nat 142:947–960

Skoczylas R (1978) Physiology of the digestive tract. In: Gans C, Gans K A (eds) Biology of the Reptilia, vol 8. Physiology B. Academic Press, London, pp 589–717

Smith M (1951) The British amphibians and reptiles (The New Naturalist). Collins, London, xiv + 318 pp

Smyth H (1962) Amphibians and their ways. Macmillan, New York, xv + 292 pp

Snyder G K, Hammerson G A (1993) Interrelationships between water economy and thermoregulation in the Canyon tree-frog *Hyla arenicolor*. J Arid Environ 25:321–329

Snyder R C (1949) Bipedal locomotion of the lizard *Basiliscus basiliscus*. Copeia 1949: 129–137

Snyder R C (1962) Adaptations for bipedal locomotion of lizards. Am Zool 2:191–203

Spellerberg I F (1973) Critical minimum temperatures of reptiles. In: Wieser W (ed) Effects of temperature on ectothermic organisms. Ecological implications and mechanisms of compensation. Springer, Berlin Heidelberg New York, pp 239–247

Špinar Z V Illustr Burian Z (1972) Life before man (Transl Shierlová M). Thames & Hudson, London, 228 pp (Reprinted 1996)

Stamps J A (1983) Sexual selection, sexual dimorphism and territoriality. In: Huey R B, Pianka R, Schoener T W (eds) Lizard ecology. Studies of a model organism. Havard University Press, Cambridge, Massachusetts, pp 169–204

Stebbins R C, Cohen N W (1995) A natural history of amphibians. Princeton University Press, Princeton, New Jersey, xvi + 316 pp

Sues H-D (1986) Locomotion and body forms in early therapsids (Dinocephalia, Gorgonopsia and Therocephalia). In: Hotton N III, Maclean P D, Roth J J, Roth E C (eds) The ecology and biology of mammal-like reptiles. Smithsonian Institute Press, Washington, pp 61–70

Sweet S S (1985) Geographic variation, convergent crypsis and mimicry in gopher snakes (*Pituophis melanoleucus*) and western rattlesnakes (*Crotalus viridis*). J Herpetol 19:55–67

Swinton W E (1970) The dinosaurs. George Allen & Unwin, London, 331 pp

Taylor D H, Guttman S I (eds) (1977) The reproductive biology of amphibians. Plenum Press, New York, x + 475 pp

Templeton J R (1970) Reptiles. In: Whittow G C (ed) Comparative physiology of thermoregulation, vol 1, Invertebrates and nonmammalian vertebrates. Academic Press, New York, pp 167–221

Thomas K R, Thomas R (1978) Locomotor activity responses in four West Indian fossorial squamates of the genera *Amphisbaena* and *Typhlops* (Reptilia, Lacertilia). J Herpetol 12: 35–41

Thorpe R S, Wüster W, Malhotra A (eds) (1997) Venomous snakes. Ecology, evolution and snakebite (Symposia of the Zoological Society of London, No 70). Clarendon Press, Oxford, xix + 278 pp

Thurow G (1976) Aggression and competition in eastern *Plethodon* (Amphibia, Urodela, Plethodontidae). J Herpetol 10:227–291

Tinsley R C (1990) The influence of parasitic infection on mating success in spadefoot toads *Scaphiopus couchii.* Am Zool 30:313–324

Tocque K, Tinsley R, Lamb T (1995) Ecological constraints on feeding and growth of *Scaphiopus couchii.* Herpetol J 5:257–265

Tracy C R (1982) Biophysical modeling in reptilian physiology and ecology. In: Gans C, Pough F H (eds) Biology of the Reptilia, vol 13. Physiology D. Physiological ecology. Academic Press, London, pp 275–321

Tracy C R, Turner J S, Huey R B (198•) A biophysical analysis of possible thermoregulatory adaptations in sailed pelycosaurs. In: Hotton N III, Maclean P D, Roth J J, Roth E C (eds) The ecology and biology of mammal-like reptiles. Smithsonian Institute Press, Washington, pp 195–206

Trompf G W (1989) Mythology, religion, art, and literature. In: Ross C A, Garnett S (eds) Crocodiles and alligators. Merehurst Press, London, pp 156–171

Turner J S, Tracy C R (1986) Body size, homeothermy and the control of heat exchange in mammal-like reptiles. In: Hotton N III, Maclean P D, Roth J J , Roth E C (eds) The ecology and biology of mammal-like reptiles. Smithsonian Institute Press, Washington, pp 185–194

Tweedle M (1977) The world of dinosaurs. Weidenfeld and Nicolson, London, 143 pp

Underwood H (1992) Endogenous rhythms. In: Gans C, Crews D (eds) Biology of the Reptilia, vol 18. Physiology E. Hormones, brain and behavior. University of Chicago Press, Chicago, pp 229–297

Underwood G (1997) An overview of venomous snake evolution. Symp Zool Soc Lond 70: 1–13

Vial J L (ed) (1973) Evolutionary biology of the anurans. University of Missouri Press, Columbia, Missouri, xii + 470 pp

Vitt L J, Pianka E R (eds) (1994) Lizard ecology. Historical and experimental perspectives. Princeton University Press, Princeton, New Jersey, xii + 403 pp

Wake M H (1977) The reproductive biology of caecilians: an evolutionary perspective. In: Taylor D H, Guttman S (eds) The reproductive biology of amphibians. Plenum Press, New York, pp 73–101

Warburg M R (1972) Water economy and thermal balance of Israeli and Australian Amphibia from xeric habitats. Symp Zool Soc Lond 31:79–111

Warburg M R (1997) Ecophysiology of amphibians inhabiting xeric environments (Adaptations of desert organisms). Springer, Berlin Heidelberg New York, xv + 182 pp

Warkentin K M (1995) Adaptive plasticity in hatching trade-offs. Proc Nat Acad Sci 92: 3507–3510

Wells K D (1977) The social behaviour of anuran amphibians. Anim Behav 25:666–693

Whittow G (ed) (1970) Comparative physiology of thermoregulation, vol 1. Invertebrates and nonmammalian vertebrates. Academic Press, New York, ix + 333 pp

Wickler W (1986) Mimicry in plants and animals (Transl Marten R D). Weidenfeld and Nicolson, London, 255 pp

Wilford J N (1985) The riddle of the dinosaur. Alfred A Knopf, New York, 304 pp

Willinston S W (1914) Water reptiles of the past and present. University of Chicago Press, Chicago, vii + 251

Wing S L, Sues H-D (1992) Mesozoic and Early Cenozoic terrestrial ecosystems. In: Behrensmeyer A K, Domulth J P, Di Michale W A, Potts R, Sues H-D, Wing S L (eds) Terrestrial ecosystems through time. Evolutionary paleoecology of terrestrial plants and animals. University of Chicago Press, Chicago, pp 327–416

Wood S C, Lenfant C J M (1976) Respiration: mechanics, control and gas exchange. In: Gans C, Dawson W R (eds) Biology of the Reptilia, vol 5. Physiology A. Academic Press, London, pp 225–274

Yeboah S (1982) Observations on territory of the rainbow lizard *Agama agama*. Afr J Ecol 20:187–192

Young J Z (1981) The life of vertebrates (3rd edn). Clarendon Press, Oxford, xv + 645 pp

Zug G R (1993) Herpetology. An introductory biology of amphibians and reptiles. Academic Press, San Diego, xiv + 527 pp

# Subject Index

Page numbers in *italics* refer to illustrations


A

Aaron's rod 222
*Acanthodactylus schmidti* 165
Acanthosauria 19
*Aclys* 66
adder (*see Vipera berus*)
advertisement 91–5
Aesop 218
aestivation 168, 170–2
*Afrixalus fornasinii* 149
*A. quadrivittatus* 90
*Agalychnis callidryas* 169
*Agama agama* (rainbow lizard) 113–14, 119–20, 134, 135, *136*, 137, 165
*A. cristata* 137, 138
*A. hispida* 32
Agamidae 196 (*see* species)
*Agkistrodon* 91
*Ahaetulla* 91
Aigialosauria 78
aktographs 160–1
Alligatoridae 210
Alligatorinae 4
alligators (*see Alligator*)
*Alligator mississippiensis* (alligator) 4, 153–4, 174–6, 203, 225
*A. sinensis* (Chinese alligator) 221
*Amblyrhynchus cristatus* (marine iguana) 82, 136–7
Ambystomidae *117*
*Ambystoma barbouri* 169
*A. texanum* 169
*A. tigrinum* 149
amnion 16
ammonia 206–7
Amphibia, evolution *25*
  decline 224–6
*Amphibolurus* 31
*A. inermis* 164–5
*A. minor* 32
*A. pictus* 32
*Amphisbaena* 56
Amphisbaenia (amphisbaenians, worm lizards) 54, 56, 205
Amphisbaenidae 54, 56, 122, 179
amplexus *142*, 143
Anapsida 19, 20, *21*, 22
*Ancylocranium* 56
*Aneides* 61, *62*
*A. lugubris* 61
Anguidae 195
*Anguis fragilis* (slow-worm) 53, 195
Ankylosauria 20, 100
*Anniella pulchra* 55, 209
*Anolis* 74, 89, 136, 137
*A. carolinensis* 89
*Anops* 56
ants, predators of 30, 32, 119–20
antagonistic muscles *8*
Anthracosauria 11, 12
Antony, Mark 222–3
Anura (frogs, toads) 19, 117–19, 124, 133–4, 139–40, 142–3, 149, 167
  shoulder girdles 46, *47*
*Apatosaurus* 22, 115
*A. ajax* *153*
apnea 83
Apoda (Gymnophiona) (caecilians) 15, 19, 29, 52, *53*, 113–14, 122, *143*
*Aporosura anchietae* 55, 184

aposematism (warning) 91–5, 119
Appalachian revolution 35
Arber, A. 8
arboreal (scansorial) adaptations 61–6
*Archaeopteryx lithographica* 38, 188
arciferous girdles 46, *47*
Archosauria 20, 22
Aristotle 154, 218
armour 100–2
arrow-poison (dart, poison) frogs (*see* Dendrobatidae, *Phyllobates*)
*Ascaphus truei* 143
Aschoff, J. 158
*Astrotia stokessii* 173
*Atractus* 97
autotomy 66, 104, 106
avian phylogeny 38

B
Bacon, F. 220
Barrie, J. M. 220
*Basiliscus* 74
*B. basiliscus* 49, *50*, 51
basking, sun 182–4
Bates, H. W. 95
Batesian mimicry 95–6, 97, 105
*Batrachoseps attenuatus* 116
batrachotoxin 151
beaded lizard (*see Heloderma horridum*)
behavioral thermoregulation 181–5
Bernard, C. 197
biological 'clocks' 157–60 (*see* rhythms)
Bible 221–2, 223–4
birds 38
*Bitis* 67
*B. peringueyi* 55, 207
*B. gabonica* (gaboon viper) *92*
*Blanus* 56
*Boa constrictor* 91
Boidae 121
*Boiga dendrophila* (mangrove snake) 218
boomslang (*see Dispholidus typus*)
*Bothrops atrox* (fer-de-lance) 126
*B. nummifer* 66–7
*Brachymeles* 146
*Brontosaurus* 22
Browning, R. 218
brumation 170
*Bufo* (toads) *47*, 117
*B. americanus* 160
*B. bufo* 74, 106
*B. calamita* (natterjack) 148
*B. fowleri* 160, 162
*B. marinus* 110, 162
*B. regularis* 160, *161*, 198, 200–1
*B. taitanus* 199
*B. viridis* 74
burrowing 52–6
Bunyan, J. 218
Byron, Lord 223, 224

C
*Cachyx* 32
*Cadea* 56
caecilians (*see* Apoda)
Caesar, Julius 222
*Caiman crocodilus* 208
*C. sclerops* 208
*Calanus finmarchicus* 2
*Calisaurus draconoides* 31
*Calloselasma rhodostoma* 127
*Calotes* 99
*C. versicolor* 89
Camptosauridae 111
*Camptosaurus* 111
cannibalism 148–9
cantharidin 214
*Captorhinus* 115
carapace 99–100, 214
carbon dioxide 198–9
*Cardioglossa gracilis* 89
*C. leucomystax* 89
*Caretta* 174
*C. caretta* 174
carnivory 115–24
Caudata (*see* Urodela)
cellulose digestion 114, 128
*Cerastes cerastes* 30, *31*, 105, 222
*C. vipera* 105, 222
Ceratopsia 100–2
Ceratopsidae 49, 100, 112
*Cerberus rynchops* 209
Chamaeleonidae (chameleons) 63–4, 89, *120*, 195, 221

Charas, M. 223
Chatterton, T. 218, 223
Chaucer 224
Chelonia (Testudinata) 4, 19, 20, 99, 112, 113, 134, 153, 188, 210, 218–19 (*see* species)
*Chelonia mydas* (green turtle) 42, 80, 113, 174, 214
*Chelydra serpentina* 214
Chelydridae 80
*Chilomeniscus cinctus* 56
Chinese alligator (*see Alligator sinensis*)
*Chionactis* 55
*C. occipitalis* 32, 56
*Chiromantis* 202
*C. rufescens* 202
*C. xerompelina* 61, 202
*Chlamydosaurus kingii* 49, *50*, 106
choanae 124
*Chrysemys* 113
*C. picta* 144, 164
*Chrysopelea* (flying snakes) 67, 68
*C. ornata* 68
chromatophores 87, *88*, 89 (*see* colour change)
chuckwalla (*see Sauromalus obesus*)
circadian rule 159, 161
classification 4, 19–22
clawed toad (*see Xenopus laevis*)
*Clemmys guttata* 164
*C. muhlenbergii* 179
Cleopatra 222–3
clutch size 102
cobras 218 (*see* Elapidae, *Naja naja*, *N. haje*, *Ophiophagus hannah*)
cocoons 199, 204–5
cold 195–6
collared lizard (*see Crotaphytus collaris*)
coloration (*see* aposematism, crypsis)
colour change 87–9 (*see* chromatophores)
*Colostethus* 119
Colubridae 121, 125, 144, 218
*Coluber viridiflavus* 67
common frog (*see Rana temporaria*)
compass orientation 175–6
competition 149–50
concealment 86–91, 162
*Conolophus pallidus* 180
*C. suberistitus* 138
conservation 225
convergence 29–33, 51, 90, 99, 111, 121
coral snakes 94, 97–8 (*see Micrurus*, *Micruroides*)
Cordylidae (zonures) 99
*Cordylurus giganteus* 208
*Corucia* 113
Cott, H. B. 89
Cotylosauria *18*, 19, 20
countershading 89
courtship 140–5
crocodile tears 220
  worship 220
crocodilians (*see* Crocodylia)
Crocodylia (alligators, crocodiles, crocodilians) 20, 22, 36, 49, 85, *86*, 112, 144, 153–4, 188, 203, 210, 214–15
  eggs 154, *155*
  locomotion 44
  food 112–13, 216–17
  territorial behaviour 135
Crocodylidae 210
*Crocodylus acutus* 82
*C. johnsoni* 217
*C. niloticus* (Nile crocodile) 85, *86*, 154, *155*, 164, 203, 217
*C. palustris* (mugger) 217
*C. porosus* (estuarine crocodile) 82, 154, 174, 216, 217
*Crossodactylus* 74
Crossopterygii (lung fishes) *12*, 14
Crotalinae (pit-vipers) 30, 121, 126, 216
*Crotalus* (rattlesnakes) 105, 172, 215
*C. adamanteus* 126
*C. atrox* *139*
*C. cerastes* 30, *31*
*C. durissus* 126
*C. viridis* 105
*Crotaphytus* 74
*C. collaris* (collared lizard) 49, 51
*C. wislizeni* 31
cryoprotectants 195
crypsis 86–91
Cryptobranchidae 142

*Cryptobranchus alleghaniensis* (hellbender) 141
*Ctenosauria pectinata* 138
*Ctenotus* 32
*C. robustus* 137
*Corucia* 113
*Cuora* 113
cycles, reproductive 161–70 (*see* phenology)
*Cyclops* 222
*Cyclorana* 204
Cynodontia 44–5
*Cynognathus* *18*, 20, 44

D
*Dactylopterus volitans* *34*
dangerous reptiles 216–17
*Dasypeltis* (egg-eating snakes) 106, 122, 153
death feigning (thanatosis) 103
defences, primary 86–91
  secondary 86, 91–5, 102–7
  structural 99–102
deflection marks 94–5
deimatic (startling) behaviour 66, 104–6
*Deinonychus anterrhopus* *48*, 49–50
*Delma* 66
*D. tincta* 66
Dendrobatidae (arrow-poison frogs) 93, 95–6, 119, 133, 150–1
*Dendrobates* 119, 150
*D. femoralis* 96
*D. galindoi* 133
dentition (*see* teeth)
*Dermochelys* 174
*D. coriacea* (leatherback turtle) 80
desiccation 202–3
*Desmognathus imitator* 96
*D. ochropaeus* 96
*Diadectes* *18*, 19, 115
Diapsida 20, *21*
*Dimetrodon* *17*, 20, 115, 180–1, 183
*D. grandis* 182–3
*D. milleri* 183
dinosaurs 22, 23, 45–6, 48–9, 95, 100–2, 110–12, 134–5, 152–6, 187–8 (*see* families, orders, species)
  extinction 35–40, 187
*Diplocaulus* *13*, 19
*Diplodactylus conspicilatus* 32
*Diplodocus* 22, 80
*Dipsosaurus dorsalis* 32, 128, 184
dispersal 172–7
*Dispholidus typus* (boomslang) 63, 106, 218
disruptive coloration 89–91
diversity, index of 1–2
Dolichosauria 78
*Draco* (flying dragons) 67–8, 136
*D. volans* *34*, *68*
*Dracunculus mediensis* 222
Dryden, J. 224
*Dryophis nasuta* 218
Dryosauridae 111

E
Eccritic temperatures *178*, 193–4
*Echis* 105
ecological analogues 29–33
ectothermy 157, 177–8, 186, 194–5
*Edaphosaurus* 180–1
eels, swimming *73*
egg-eating snakes (*see Dasypeltis*)
eggs 16, 24, 27, 37, 102, 131, 149–51, *152*, 153–6, 188, 189, 206 (*see* oviposition)
  size 152
Egyptian cobra (*see Naja haje*)
Elapidae 125, 127, 164, 218
Elasmosauridae 75
*Elasmosaurus* 75, *76*
*Eleutherodactylus* 199
*E. cundelli* 151–2
*E. gaigeae* 95–6
*Elodea* 110
*Elosia* 74
*Emydocephalus annulatus* *82*
Emydidae 41, 80, 113
*Emys orbicularis* 3, 4
endothermy 157, 177, 186
*Ensatina escholtzii* 96
enzymes 125
epigamic coloration 95
*Eremias namaquensis* 95
*Eretmochelys* 174
*E. imbricata* (hawksbill turtle) 80, 81, 214

*Eryops* *14*, 19, 115
*Erythrolamprus* 97
*Eryx conicus* 56
*E. johni* 56
*Eumeces* 74
excretion 206–7
*Exocoetus* 34
extinction 33–40, 224–5
eyes 56, 80, 89–90, 90–1, 119, 162, 205–6

F
Fangs 103 (*see* teeth)
feathers 188
fer-de-lance (*see Bothrops atrox*)
fertilization 141–5
fighting 132–8, *139*
firmisternal girdles 46, *47*
flight 67–72
flying snakes (*see Chrysopelea*)
foam nests 150–1
frogs 5, *7*, 46–7, 124–5 (*see* species)
  jumping 6, 46–8, 66–7
  legs, eating 213–14
  vocalization 104, 138–40
frothing (salivating), tortoise 189–90, 193

G
Gaboon viper (*see Bitis gabonica*)
*Galeopithecus volans* *34*
*Gallimimus* 48
gametes 131
garter snakes (*see Thamnophis*)
gastroliths 77, 110, 114
*Gastropyxis smaragdina* 91
gastrosteges 82
*Gastrotheca* (marsupial frogs) 150
*G. marsupiata* 27, 150
*G. monticola* 86–7
*Gavialis gangeticus* (gharial) 112–13
geckos (*see* Gekkonidae)
*Geckonia chazaliae* 32
*Gehyra variegata* 182, *183*
Gekkonidae (geckos) 64, *65*, 182
  (*see* species)
*Geochelone pardalis* 193
*G. sulcata* 144, 162, *163*, 189, 207
*Geococcyx* (road runners) 85
geosaurs (*see* Thalattosuchia)
Gerrhonotidae 195
gharial (*see Gavialis gangeticus*)
Gila monster (*see Heloderma suspectum*)
glass lizards (*see Ophisaurus*)
gliding 67–72
glottis, protrusible 124
*Glyptosaurus giganteus* 53
gonadotropin-releasing hormone
  (GnRH) 167
*Gopherus agassizii* 203
*G. berlandieri* 171
*G. polyphemus* 162–3
*Graptemys* 80, 134
grass snake (*see Natrix natrix*)
green turtle (*see Chelonia mydas*)
'greenhouse' effect 37
Griffiths, R. A. 148
ground substance 202
guanine 87, 206 (*see* excretion)
Gymnophiona (*see* Apoda)

H
*Habrosaurus* 73
Hadrosauridae 112, 134–5, 187–8
Halberg, F. 158
*Haptodus* 183
*Hardella* 113
Harris, V. 135
hellbender (*see Cryptobranchus*
  *allaghaniensis*)
Helodermatidae 93–4, 125, 195
  (*see Heloderma*)
*Heloderma* 218
*H. horridum* (beaded lizard) 93
*H. suspectum* (Gila monster) *93*, 216
*Hemiergis decresiensis* 165
*Hemiphractus* 150
*H. bubalis* *150*
*Henodus* 77, *78*
herbivory 109–15, 195
Herodotus 220
*Heterodon nasicus* 103
hibernation 165, 168, 170–2
hissing 105 (*see* vocalization)
*Holbrookie maculata* 184
homeothermy (homoiothermy) 174, 180

*Homopus signatus* 179
horned lizards (*see Phrynosoma*)
human food 213–15
Hydrophiidae (sea snakes) 2, 82–3, 125
Hydrophiinae 82
*Hydrosaurus* 74
*H. pustulosus* *45*, 74
*Hynobius nebulosus* 149
*Hyla* 74
*H. arborea* 67, 162
*H. arenicola* 181–2
*H. cinerea* 62
*H. crucifer* 62
*H. faber* 133
*H. femoralis* 62
*H. leucophyllata* 90
*H. regilla* 89, 224
*H. savignyi* 89
*H. squirilla* 62
*H. truncata* 110
*H. vasta* 62
*H. venulosa* 67
*H. versicolor* 104
Hylidae (tree frogs) 150 (*see* species)
*Hylodes longirostris* 90
Hynobiidae 142
hyoid 15, 117–18
*Hyperolius marmoratus* *87*
*H. viridiflavus* 202
Hypselophodontidae 111

I
Ichthyopterygia 19, 20
Ichthyosauridae (ichthyosaurs) 19, 20, 74
*Ichthyosaurus* *75*
*Ichthyostega* 11, *12*
Ichthyostegalia 11, 19, 72–3
*Iguana iguana* 51, 137, 138, 152
Iguanidae 195 (*see* species)
*Iguanodon* 22, 112, 115
incubation 153
Indian cobra (*see Naja naja*)
iridium anomaly 36, 39
iridiophores (guanophores) 87, *88*, 202

J
Jacobson's organ 121
jaw disarticulation *123*, 125
Jepson, G. 35
jugular shunts 191–2
jumping adaptations, frogs 6, 46–8, 66–7
  lizards 46
  snakes 66–7

K
K-T boundary 35–6, 37, 39
keratin 102
king cobra (*see Ophiophagus hannah*)
*Kinosternon baurii* 179
*Kinixys spekii* 193
*Kronosaurus queenslandicus* 75, *76*
*Kuehneosaurus* 68, *69*

L
Labyrinthodontia 11, 13, *14*, 19
*Lacerta* 74
*L. agilis* 173
*L. saxicola* 144–5
*L. vivipara* 195
Lacertidae 137 (*see* species)
Lack, D. 132
Lamarre-Picquot, P. 189
lamellae 65
*Lampropeltis* 97
*Lampropholis guichenoti* 165
Lanthonotidae 125
Laramide revolution 35, 36
*Laticauda* 173
Laticaudinae 82
leatherback turtle (*see Dermochelys coriacea*)
*Leiolepis* 22
leopard frog (*see Rana pipiens*)
*Lepidochelys* 174
Lepidosauria 19, 20, 22
Lepospondyli 14, 19
*Leposternon* 56
*Leptodactylus* 150
*Leptopelis bocagei* 204
*Leptopterygius acutirostris* 77
*Lerista* 31–2
*Limnodynastes spenceri* 204
Lissamphibia 15, 19, 21
*Lithodytes lineatus* 96

lizards (*see* Squamata, species)
locomotion, aquatic 72–83
  arboreal 67–72
  bipedal 46–51
  concertina 57–8
  limbless 56–60
  quadrupedal 41–6
  rectilinear 54, 58
  serpentine 57, *58*
  sidewinding 58, *59*
lung fishes (*see* Crossopterygii)
*Lygosoma laterale* 170
*L. sundevalli* 55–6
lyre snakes (*see Trimorphodon*)
*Lytta* 214

M
*Mabuya* 89
*M. quinquetaeniata* 95, 165, *166*, 167, 203
*Malaclemys* 134
*M. terrapin* *211*, 214
*Malpolon monspessulanus* (Montpellier snake) 204
mammal-like reptiles (*see* Therapsida)
mangrove snake (*see Boiga dendrophila*)
marine iguana (*see Amblyrynchus cristatus*)
marking 172, 174
marsupial frogs (*see Gastrotheca*)
Mayhew, B. 204
medicine 215–16
meiosis 131
melanophores 87, *88*, 89
*Melanosuchus niger* 210
Mertensian mimicry 96–8
*Mesalina lugubris* 32
Mesosauria 78, *79*
*Mesosaurus* 78, *79*
metabolism 32–3, 127–9, 171
Microsauria 13
*Micruroides* (coral snakes) 94
*Micrurus* (coral snakes) 94, 97
  *M. diastema* *98*
  *M. elegans* 98
midwife toad (*see Rhinoderma darwinii*)
migration 172–7
mimicry 95–8
  Batesian 95–6, 97, 105
  Mertensian 96–8
  Müllerian 98, 105
mitosis 131
*Moloch horridus* (thorny devil) *30*, 32, 119–20, 208
monitors (*see* Varanidae)
*Monopeltis* 56
monophyly 11, 15
Montpellier snake (*see Malpolon monspessulanus*)
Mosasauria (mosasaurs) 78, *79*
Moses 222
moulting 99
mugger (*see Crocodylus palustris*)
Müller, F. 96
Müllerian mimicry 98
muscles, antagonistic *8*
mythology 54, 218–24

N
*Naja haje* (Egyptian cobra) 222
*N. naja* (Indian cobra) *103*
nasal glands 204
*Natrix natrix* (grass snake) 3, 105, 106–7, 195
natterjack (*see Bufo calamita*)
navigation 174–6
Nectridea 12
*Nectophrynoides* 143
*Necturus maculosus* 206
neoteny 149, 196
*Nereis* 57
newts *43* (*see Taricha*, *Triton*, *Triturus*)
Nile crocodile (*see Crocodylus niloticus*)
nomenclature 4
nostrils 12, 55–6, 124
Nothosauria 19, 75, 77
*Notophthalmus* 4, 97
*N. viridescens* 97, 175
*Nucras tessellata* 32

O
*Ocadia* 113
*Oedipus* 61
omnivory 115–22
oophagy 149
*Ophiacodon* 115

*Ophidiocephalus* 66
*Ophiomorus punctatissimus* 54
*Ophiophagus hannah* (king cobra) 144, 154
*Ophisaurus* (glass lizards) 54
*O. apodus* 53
Ornithischia 20, 22, *23*, 152–3
*Ornithomimus* 48
Ornithopoda 20, 112
osmosis 200, 209–10
osteoderms 99, 100
*Osteolaemus osborni* 179
*Otocrypsis* 69
oviposition 145–8
*Oviraptor* 153, *154*
oviparity 146–8
ovoviviparity (viviparity) 146–8, 160, 195–6
*Oxybelis* (vine snakes) 91, 106
*Oxyuranus scutellatus* (taipan) 126

**P**
pH 198–9
*Pachycephalosaurus grangeri* 135
*Palaeosuchus palpebrosus* 179
*Palmatogecko rangei* 52
Pangaea 36
parallel evolution (*see convergence*)
Parapsida 20, *21*
parental care 145, 147, 149–56
parthenogenesis 144–5
*Pelamis platurus* (yellow-bellied sea snake) 2, 92, 173–4
*Pelodytes punctatus* 74
*Pelomedusa subrufa* 144
pelvis 23
Pelycosauria *17*, 19, 20, 179, 180–1, 182–3, 187–8
perentie (*see Varanus giganteus*)
*Periophthalmus* 61
peritoneum 89
*Phagocotus gracilis* 169
phenology (seasonal cycles) 165, 167–70
pheromones 141, 170
photoperiodism 160, 167–8
*Phrynocephalus* 51
*P. mystaceus* 32, 66
*Phrynosoma* 30, 32
*P. coronatum* 191, *192*, *193*
*P. platyrhinos* *30*, 55
*Phyllobates* (arrow-poison frogs) 96, 150
*P. aurotaenia* 96
*P. lugubris* 95–6
*Phyllomedusa* 61, 145, 202
*P. ihrenegi* 202
*P. sauvagei* 202
*Physalaemus* 150
*P. nattereri* *94*
*P. pustulosus* 150
*Physignathus lesuerii* 49
physiological thermoregulation 180, 185–95
*Pipa pipa* (Surinam toad) 27, 74
Pipidae 118, 145
pit vipers (*see* Crotalinae)
pits, sensory 30, 121
*Pituophis* 105
*P. melanoleucus* 105
*Pityas nigromarginatus* 67
*Placochelys* 77
Placodontia 77, *78*
*Plateosaurus* 115
Platynota 78
plesiosaurs 19, 20, 75, 80 (*see* species)
Plethodontidae *117*, 142, 196 (*see* species)
*Plethodon* 62, 133
*P. cinereus* (red-backed salamander) 96, 116, 133
*P. hubrichtii* 96
*P. jordani* 96
*Pleuroderma brachyops* 94
Pliny 154, 218
*Pliocerus elapoides* *98*
pliosaurs 19, 75, 77
poikilothermy 177, 181
poison frogs (*see* arrow-poison frogs)
*Polypedates leucomystax* 89
Pope, A. 218
populations, declining 224–5
predators 85–6, 94, 150–2
*Prostherapis panamaensis* 133
*Protobatrachus* 15
*Protoceratops andrewsi* *152*
Protorosauria 77
*Protosuchus* 49
*Protothica* 148

*Pseudemys* 80, 113
*P. scripta* 144
*Pseudoboa* 97
*Pseudotriton montanus* 97
*P. ruber* 97
Psittacosauria 49
*Psyllophryne didactyla* 119
*Ptenopus garrulus* 140
*Pteranodon* 70
Pterodactyloidea 20, 71
Pterosauria 20, 22, 69–72
*Ptychozoön* *34*, 67–8
Pygopodidae (snake lizards) 66, 104
*Pygopus nigriceps* 32
*Python molurus* 91, 128, 189, *190*, 215
*Pyxicephalus adspersus* 199, 204

Q
*Quetzalcoatlus* 70

R
r-K selection 27, 187
*Rana* 89, 224
*R. aurora* 224
*R. muscosa* 224
*R. pipiens* (leopard frog) 86, 181, 185
*R. sylvatica* (wood frog) 148, 195
*R. temporaria* (common frog) 67, 89, 148
*R. utricularia* 148
rattlesnakes (*see Crotalus*)
red-backed salamander (*see Plethodon cinereus*)
Redi, F. 223
Reptilia, evolution *26*
respiration 15, 83, 106, 124, 198–9, 205–7
retrahescence 170
road runners (*see Geococcyx*)
*Rhacophorus* 150
*R. fasciatus* 90
*R. nigropalmatus* 67
*R. reinhardtii* *34*
Rhamphorhynchoidea 71–2
*Rhamphorhynchus* *70*, 71
*R. phyllurus* 71
*Rhineura floridana* 209
*Rhinobothrium lentignosum* 97
Rhinodermatidae 151
*Rhinoderma* 151
*R. darwinii* (midwife toad) 151
Rhipidistia 15
Rhynchocephalia 20, 22
*Rhynchoedura* 32
rhythms (*see* biological clocks, cycles, reproductive)
  circadian 158, 182
  circannual 158, 167
  circalunar 158, 167
  diel 165
  endogenous 158, *159*, *161*
  exogenous 158
Riedesel, ML (B) 190
Romer, A. S. 78

S
*Sagittarius serpentarius* (secretary bird) 85
*Salamandra* 125
*S. maculosa* (fire salamander) 93, 219
*S. salamandra* 149
salamanders (*see* Urodela)
salivary glands 125, *126*
salt-excreting glands 210, *211*
sand running 51–2
  swimming 54
Saurischia 20, 22, *23*
*Sator grandaevus* 184
*Sauromalus* 32, 114
*S. obesus* (chuckwalla) 138, 171
*S. varius* 138
*Sauropleura* 13
Sauropoda 20, 22, 110, 114
scansorial (climbing) adaptations 61–6
*Scaphiopus* (spadefoot toads) 54–5, 168, 210
*S. couchi* 204
*Sceloporus* 99
Schultz, C. 214
Scincidae 113, 195 (*see* species)
*S. ocellatus* 55
*S. officinalis* 55
*Scolosaurus cutleri* 100, *101*
sea grasses (*see* turtle grasses)
sea snakes (*see* Hydrophiidae, *Pelamis platurus*)
seasonal cycles (*see* phenology)

secretary bird (*see Sagittarius serpentarius*)
Serpentes (snakes) 147, 204
feeding 121–4, 125–7
fighting 138, *139*
locomotion 57–70
sidewinding *59*, 60
size 121
venoms 102, 125–7, 216–18
setae 65
sex determination 188
*Seymouria* 16
shadow elimination 89
Shannon-Wiener index 1
shape 23
Shakespeare 8, 218–19, 220, 221, 222–4,
*Sharovipteryx mirabilis* 69
shell (*see* carapace)
shuttling 181, 184
*Sinosauropteryx prima* 188
*Siren* 110
*S. intermedia* 204
Sirenidae 72–4 (*see* species)
size 23, 121, 134, 136, 178–81
and thermoregulation 178–81
sloughing 99
slow-worm (*see Anguis fragilis*)
*Smilisca baudinii* 204
*Sminthillus limbatus* 199
Smith, M. 5
snake lizards (*see* Pygopodidae)
snakes (*see* Serpentes)
solenoglyphs 123
*Sordes pilosus* *188*
spadefoot toads (*see Scaphiopus*)
'spectacles' 56, 205
Spenser, E. 218
spermatophores 131, 141–2, 143
*Sphaerodactylus cinereus* 65
*S. lineolatus* 182
*Sphenodon* 20, *22*, 179
*S. punctatus* (tuatara) 104, 128, 134, 179
Sphenodontidae 111
*Sphenomorphus kosciuskoi* 137
*S. quoyi* 137
*Sphenops* 54
*S. chalcides* 54
*S. sepsoides* 55
spines 99, 100–2
Squamata 20, 128, 147, 154–5, 188, 194–5, 207 (*see* families, species)
startling behaviour (*see* deimatic behaviour)
Stegocephalia 162
*Stegoceras validus* *135*, 145
Stegosauria 20, 111
*Stegosaurus armatus* *101*, 179
*Stenodactylus* 51
*Sternotherus odoratus* (stinkpot) 184
stinkpot (*see Stegotherus odoratus*)
*Storeria* 144
Strabo 220
*Stretosurus macromeros* 75
stridulation 104, 105
structural defences 99–102
*Struthiomimus* 46
Suckling, J. 220
Surinam toad (*see Pipa pipa*)
swimming (*see* locomotion, aquatic)
Synapsida 20, *21*
Synaptosauria 19, 20, 77
synchronizers 158

T
*Tachyenemis fornasinii* *90*
tadpoles 109–10, 116, 169
taipan (*see Oxyuranus scutellatus*)
*Tarentola annularis* 165, 166, 167, *203*
*Taricha* 4
*T. granulosa* 97
*T. torosa* 96
Teiidae 148, 195
teeth 103–4, 110–12, 116–17, 122
Temnospondyli 11, 19
temperature, lethal 174
Tennyson, A Lord 220
*Tenontosaurus* 112
*Teratoscincus scincus* *45*
*Terrapene carolina* 175
*T. ornata* 170, 190
territory 132–8, 151
Testudinata (*see* Chelonia)
Testudinidae (tortoises) 3, 4, 36, 107, 113 (*see* species)

*Testudo graeca* 190
*T. leithii* 207
*Thalassia* 81
Thalattosuchia (geosaurs) 78
*Thamnophis* (garter snakes) 144, 172
*T. sirtalis* 97, 195
thanatosis (death feigning) 103
*Thecadactylus rapicaudus* 182
Thecodontia 20, 22
*Thelotornis kirtlandii* (twig snake) 63, *64*, 91, 106
Therapsida (mammal-like reptiles) 17, *18*, 19, 20, 44–6, 146–7, 180, 181, 185–6
thermoregulation 147–8, 165, 178–81
behavioral 181–5
physiological 185–8
Theromorpha 20
Theropoda 20, 22, 35, 38, 49
thorny devil (*see Moloch horridus*)
*Tiliqua rugosa* 95
Titanosauridae 112
toads 44, 45–6 (*see* species)
toe fringes 44, *45*
pads 61, *62*
webs 74
tongues *117*, *118*, 119, *120*, 121–2, 124
*Torosaurus latus* *100*
tortoises (*see* Testudinidae)
toxins 119, 150, 151
*Trachodon* 80
transpiration 62, 80, 157, 178, 197–204, 208
*Trimeresaurus albolabris* 62, *63*
*Trimorphodon* (lyre snakes) 218
Trionychidae 80
*Triton* (newts) 145
*Triturus* (newts) 4, 73, 125, 145, 162, 206–7
*T. vulgaris* 141, 171
tuataras (*see Sphenodon*)
turtle grasses (sea grasses) 81, 113, 174
twig snake (*see Thelotornis kirtlandei*)
*Tylosaurus* 78, *79*
*Typhlops* 179
*Typhlosaurus* 31, 32
*Tyrannosaurus rex* 22, 49, *50*

U
Ultra violet (UV) radiation 89, 225
*Uma* 52, 55
*U. notata* *45*, 51
urea 206–7, 210
uric acid 206–7, 210
urine 107
Urodela (Caudata) 3, 4, 19, 41, 42, *43*, 61, 73–4, 141–2, 145, 149, 167 (*see* species)
*Uromacer* (vine snakes) 91
*Uromastyx* 32, 99, 113, 215
*U. hardwickii* 208
*U. microlepis* 102
*Uta stansburiana* 203

V
Varanidae (monitors) 125, 148, 195
*Varanus* 119, 137
*V. bengalensis* *137*
*V. eremias* 31
*V. exanthematicus* 103, 171
*V. giganteus* (perentie) 51
*V. niloticus* 171, 209
*Velociraptor* 48
venom 102–3, 124–7, 218
glands *126*
vine snakes 91 (*see Oxybelis*, *Uromacer*)
*Vipera* 91
*V. berus* (adder) 3, 105, 126, 169, 172, 195
Viperidae 121, *123*, 125, 126, 127, 218 (*see* species)
Viperinae 126
Virgil 223
viviparity (*see* ovoviparity)
vocalization 104, 138–40, 144, 151, 172
vomerine teeth 125

W
Wallace, A. R. 67
warning coloration (*see* aposematism)
sounds 104, 105
water loss (*see* excretion, respiration, transpiration)
recolonization 211–12
uptake 207–9
Wesley, J. 224

White, G. 219
wood frog (*see Rana sylvatica*)
worm lizards (*see* Amphisbaenia)
Wright, B. 217

**X**

Xanthophores 87, *88*
*Xantusia* 205
*Xenopus laevis* (clawed toad) 74, 206

**Y**

Yellow-bellied sea snake (*see Pelamis platurus*)

**Z**

*Zeitgeber* 158
zonures (*see* Cordylidae)
*Zostera* 81
*Zygaspis* 56

Printing: Saladruck, Berlin
Binding: Buchbinderei Lüderitz & Bauer, Berlin